INFLECTION POINT

How the Transition to Renewable Energy Became Inevitable

ROD PERRY

ISBN: 978-0-9905465-1-1
Published by Rod Perry

Book Design by Rudy Ramos

For my family, from which all good
things in my life flow

Contents

PART I

RENEWABLES NOW WIN ON COST ALONE

CHAPTER 1

The Year Everything Changed

At the start of the year 2025, the world was engulfed in drama; from active "hot" wars to populist political upheaval and economic strains. Against this tumult, few people were considering the technology behind how their electricity was generated. Yet in the quiet accumulation of data points, in the steady hum of wind turbines and the silent collection of photons on solar panels across every continent, something extraordinary happened. Humanity crossed a threshold that historians will mark as one of the most consequential turning points in our species' relationship with energy.

For the first time in the industrial age, clean energy sources accounted for almost all the new electrical capacity created that year.[1] Not in a single country. Not in a progressive state or a wealthy region. Globally. The implications of this quiet milestone cannot be overstated. The generation of electricity by fossil fuels had peaked and begun its structural, permanent decline. This wasn't supposed to happen quite yet. Not in a world where political headwinds blew against climate action, where fossil fuel subsidies still dwarfed renewable energy support, and where powerful interests fought tooth and nail to preserve the status quo.[2] But it happened anyway, and that's precisely what makes 2025 the inflection point from which there is no return for fossil fuels.

If you are even a half interested student of history, you know we have seen these inflection points before. In 2000, Kodak was a healthy business concern dominating the film photography business. Digital cameras had been around for a decade but were dismissed as toys that could never compete with real film cameras, but by 2012, Kodak was bankrupt. The same is true in the conversion from Blockbuster's VHS tapes to Netflix streaming. From mainframes to smartphones. From wired to wireless. Often even industry pundits and business experts do not recognize the inflection point when a legacy technology becomes a "dead man walking".

The Quiet Revolution

Understanding why 2025 marked an irreversible turning point requires understanding what changed and what didn't. Climate change continued to accelerate. Political polarization around energy policy remained fierce in many countries. Fossil fuel companies still wielded enormous political and financial influence. What changed was something far more powerful than policy or politics: the fundamental economics of electrical generation. Simply put, renewables became the least expensive way to generate electricity for many countries around the world.

By mid-2025, solar power had become 41 percent cheaper than the lowest-cost fossil fuel option globally. Onshore wind power was 53 percent cheaper.[3] These aren't projections or best-case scenarios, but market realities reflected in actual power purchase agreements being signed across the world. When a utility executive in Texas or China, or a rural cooperative manager in Kenya evaluates their options for new generating capacity, the spreadsheet increasingly points to one clear indisputable winner: Renewables.

This economic reality is not a one-time event, but a self-reinforcing cycle that, once begun, spins faster and grows larger. Every gigawatt of solar installed drives manufacturing scale, which reduces costs further, which makes solar more competitive, which drives more installations. This isn't a theory, but the same manufacturing principle that drove the cost of computers from dozens of room-filling mainframes costing millions to billions of smartphones costing hundreds.

For more than four decades, solar panel prices have declined by approximately 20 percent with each doubling of global cumulative installed capacity. In the last decade, solar costs have plummeted by roughly 90 percent, while wind turbine costs have fallen 70 percent over the same period.[4] These aren't gradual improvements. These are exponential cost improvements that leave fossil fuels further behind with each passing year.

Four Converging Forces

Four powerful forces converged in 2025 to create this inflection point, each reinforcing the others in ways that make the transition to renewable energy both inevitable and accelerating.

First, renewable energy technology matured. The "holy grail" of renewable energy proved to be affordable energy storage in the form of batteries, which arrived not as a lab breakthrough but as the result of tens of millions of electric vehicles being manufactured globally. The more you make of something, the lower the price becomes. Battery costs fell 90 percent from 2010 to 2024, making grid-scale storage economically competitive for the first time.[4] Renewables have an intermittency problem (the sun only shines during the day, the wind doesn't always blow, etc.) that had plagued them for decades. The solution didn't come from better engineering or design, but by market-driven manufacturing scale. Tesla and BYD weren't trying to change the electric grid when they ramped up battery production; they were building cars. Affordable grid-scale electrical storage was a happy byproduct.

Second, grid management technologies evolved from crude on-off switches to AI-powered systems capable of coordinating thousands of distributed generation sources and subscribers with microsecond precision. Thousands of home solar installs, backed by in-home batteries, can now be joined together to create dynamic Virtual Power Plants. The "baseload power" arguments that fossil fuel advocates had wielded for years lost their merit, because power would keep flowing for hours in the dark and on days without wind. Modern grids demonstrated conclusively that reliability improves with the addition of renewable technology, not despite it.

Third, economics shifted decisively. This is no longer about subsidies or carbon taxes making renewables artificially competitive. In most markets around the world in 2025, renewables are simply cheaper, period. Unsubsidized. Head-to-head. The levelized cost of electricity (LCOE) from new solar projects regularly comes in at less than half the cost of new coal or gas plants, and potentially less than the operating cost alone of the least efficient fossil fuel plants.[5] This cost advantage isn't trivial. It is overwhelming and it is increasing. As fossil fuel costs remain relatively constant (or increase due to depletion of easily accessed reserves), renewable costs continue to rapidly decline. Every percentage point that solar costs fall makes fossil fuel power generation less competitive, limiting access to capital for fossil fuel extractors, stranding more assets (oil and coal in the ground that is too expensive to pull out), reducing industry revenues, and weakening their ability to compete. It is an economic death spiral for fossil fuels and a virtuous, upward cycle for renewables because extraction industries cannot match manufacturing efficiencies at scale.

Fourth, markets took over from politics. Perhaps most importantly, the renewable energy transition stopped depending primarily on government policy and started being driven by corporate procurement and investor decisions. When Amazon, Apple and Microsoft commit to 100 percent renewable energy, they are not making political statements.[6,7,8] They are making economic calculations that serve their own interests.

This shift from politics to markets is what makes the 2025 inflection point irreversible. Government policies can change with elections, but market fundamentals, once shifted, are much harder to reverse. No amount of political rhetoric can make the sun more expensive or fossil fuels substantially cheaper and no subsidy for coal can overcome a 200% cost disadvantage.[5] No regulatory rollback can force utilities to choose the more expensive option when their shareholders demand profits and their customers demand affordable electricity in the long run.

Why This Time Is Different

Every generation that has watched the renewable energy movement has heard promises before. The 1970's oil crisis sparked enthusiasm for solar that faded when oil prices dropped. The 1990's brought renewed hope around climate policy that floundered on resistance from fossil fuel companies and their minions in government. The 2000's saw the rise and fall of "cleantech" as a venture capital darling, leaving investors burned when companies couldn't compete with the wave of cheap natural gas newly produced from fracking shale deposits.

The answer to why 2025 is different lies in understanding what fundamentally changed between those earlier moments and now. Previous "green waves" were driven primarily by environmental concern, energy security fears, or policy support. All these motivators are vulnerable to changing circumstances. Oil prices stabilize, and concern about energy security fades. Political winds shift, and subsidies disappear. Economic recession hits, and environmental priorities take a back seat to jobs and growth. And capital for these risky ventures remains scarce. But 2025's inflection point is driven by something more fundamental and dearer to humans. Low cost. Solar technology now follows exponential cost-decline curves that fossil fuels cannot match.[9] This isn't about tree hugging, saving the earth or anything "green". It is about the inevitable financial result of learning curves applied to scalable manufacturing technologies.

In 1936, Theodore Paul Wright, an aeronautical engineer working in the American aircraft industry, noticed something remarkable. Every time his company doubled its cumulative production of aircraft, the labor hours required to build each plane fell by approximately 15 percent. What Wright had discovered was a mathematical principle that would eventually determine the fate of entire industries, reshape global geopolitics, and make the renewable energy revolution inevitable. Today, it is known as Wright's Law, or the learning curve of industry and it is the reason fossil fuels are doomed as a source of energy for many applications.[10]

Wright's Law is remarkably simple. For every cumulative doubling of units produced, costs decline by a constant percentage. For solar panels, that percentage has held remarkably steady at about 20 percent for over four decades. This means that as we move from gigawatts to terawatts of installed solar capacity (a doubling we're accomplishing in increasingly short time periods) costs will continue to fall with each doubling.[4] Fossil fuel extraction and combustion, by contrast, remain flat or become more expensive. The easiest-to-access deposits were extracted first, and those fuels were burned decades ago. Over time, remaining reserves require deeper drilling, more extreme operating environments, more complex extraction techniques that all push costs upward. While renewables get cheaper with scale, fossil fuels will inevitably get more expensive because they are a finite resource.

The Evidence Mounts

By 2025, the evidence of an energy inflection point has become obvious. Globally, over 90 percent of new electricity generating capacity added in 2024 came from renewable sources. In absolute terms, more solar capacity was installed in 2024 than all other electricity sources combined, an amount of new generation equal to more than 500 new nuclear plants.[1] In 2024, 92% of new electrical production in oil drenched Texas was renewable energy.[11] But perhaps more telling than installation figures was what happened to electricity generation itself. For decades, renewable growth simply met the ever-increasing demand for electricity, with fossil fuels maintaining their dominant share in parallel. In 2025, that changed. Clean energy sources grew fast enough to meet all new demand plus begin displacing existing fossil fuel generation. Coal plants began shutting down not because regulations forced them to, but because they were no longer economically viable.[12]

Financial markets reflected this shift with remarkable clarity. Investment in renewable energy and storage exceeded $2.4 trillion globally in 2024, roughly double the investment in fossil fuel production and infrastructure. More striking still, much of these investments weren't driven by or tied to government mandates or subsidies. Private capital, seeking the highest, safest returns, recognized the inevitable direction of energy markets and positioned accordingly. For example, Climate First Bank, a bank that only loans to renewable energy projects, is not only operating, but is a very successful billion-dollar business.[13]

The Irreversibility Principle

What makes an inflection point truly irreversible? For energy systems, it is when the momentum of change creates self-reinforcing cycles that no single actor, not even a superpower government, can stop. Consider the forces now in motion. Every additional gigawatt of solar installed drives manufacturing scale and pushes costs down further. Lower costs make solar more competitive in more markets, driving additional installations. This creates a manufacturing cost curve accelerator that feeds on itself. For fossil fuels to compete, they would need their costs to fall at similar rates. But oil, gas, and coal aren't manufactured, they are extracted and it is impossible to improve their costs in a meaningful manner. Similarly, every electric vehicle sold adds to the installed base of mobile battery storage and every grid storage battery makes EVs less expensive. This cycle drives down battery costs through manufacturing scale, making both EVs and grid storage cheaper, which increases adoption further. It is Wright's Law on a global scale.

Perhaps most importantly, a new generation of engineers, entrepreneurs, financiers, and policymakers is being trained in renewable energy systems. Universities graduate thousands of solar engineers each year. Startups pivot toward battery technology and grid software. Manufacturers develop electric boats, jet skis and airplanes.

Investment analysts develop deep expertise in clean energy project finance. This human capital doesn't evaporate with the changing political winds, as it is a permanent cultural shift in which talent and ambition grow.

What's Ahead in Your Reading

This book will explore how we reached this inflection point and where the renewable revolution leads next. We'll examine the global landscape of renewable deployment, from China's manufacturing juggernaut to Europe's grid integration lessons to the Global South's leapfrogging development model.

We'll dive deep into the technologies that made this transformation possible. Not just solar panels and wind turbines, but the storage systems, smart grids, virtual power plants and digital management tools that turned intermittent renewable power into reliable, dispatchable electricity.

We'll confront the political headwinds directly, examining why fossil fuel interests fight so fiercely despite deteriorating economics, how regional variations in politics affect deployment speed, and why, ultimately, politics cannot stop a transition driven by market fundamentals.

And finally, we'll explore what a fully electrified world looks like. Not as utopian fantasy, but as the logical endpoint of current trajectories. A world where transportation, heating, industry, and agriculture all run on clean, abundant electricity. A world where air pollution deaths decline dramatically, where true energy independence reduces geopolitical tensions, where climate stabilization becomes possible.

This is not a book about what we should do or could do. It is a book about what is happening and why it cannot be stopped. The renewable energy revolution is not something to aspire to; it is now inevitable.

The year 2025 didn't announce itself with fanfare. But decades from now, historians will mark it as the year humanity turned the corner on the greatest energy transformation in human history. Not because we finally decided to act on climate change, but because clean energy became cheaper than dirty energy, and markets did what markets do. They followed the money toward the future.

CHAPTER 2

The Math That Changed the World

The principle of Wright's Law is elegantly simple: for every doubling of cumulative production of a manufactured good, costs decline by a constant percentage. That percentage varies by technology. Wright found 15 percent for aircraft, but the exact rate for other technologies depends on the complexity of the product and the potential for manufacturing optimization. What makes this principle so powerful is not just that it happens, but that it happens predictably, repeatedly, and apparently without limit for manufactured goods. Double production, costs fall. Double it again, costs fall again. This isn't magic. It is the accumulated result of thousands of small improvements: better tools, refined processes, improved materials, worker expertise, supply chain optimization, and design innovations.

For solar photovoltaics, the price of solar panels has declined by 20 percent with each doubling of global cumulative capacity for more than four decades. This means that in the last forty years, the cost of solar panels has dropped 99%.[14] In a similar fashion, lithium-ion battery cells have seen prices fall by around 97 percent since 1991, with an average reduction of 19 percent for every doubling of capacity, and this rate does not yet appear to be slowing down.[15] This is why 2025 marks an inflection point. We've reached the part of the exponential curve where renewable energy costs have fallen so far below fossil fuels that market forces take over from policy interventions. The dollar signs simply overwhelm all other considerations.

Why Fossil Fuels Cannot Compete

It is key to understand that Wright's Law only applies to manufactured goods, not extracted resources. You cannot "manufacture" oil or coal. You extract them from the ground. And the extraction process follows the opposite trajectory from manufacturing. Geologists and petroleum engineers find the easiest-to-access deposits first. The oil that gushes from the ground under natural pressure. The coal seams near the surface. The natural gas in conventional reservoirs. But over time, these reserves deplete. What remains requires more sophisticated techniques, deeper drilling, more hostile environments, more energy input per unit of energy extracted. Offshore platforms in ever-deeper water. Hydraulic fracturing to extract gas from tight shale formations. Tar sands that require massive energy inputs to process. Arctic drilling in extreme conditions. These efforts do not reflect abundance, they reflect depletion.

Each of these advances in extraction technology is impressive from an engineering standpoint. But from an economic standpoint, they are moving in the wrong direction. Costs increase with extraction complexity. The learning curve for fossil fuels is inverted or more accurately, they don't have a learning curve at all. They have a depletion curve and after a hundred years of consuming fossil fuels, we are deep into the depletion curve. Meanwhile, every solar panel manufactured makes the next one cheaper. Every wind turbine installed makes the next one more economically attractive. Every battery pack produced drives down the cost of the next. This creates a scissors graph where the lines inevitably cross. And in 2025, they've crossed decisively, globally, and irreversibly.

The Crossover Point

By 2024, the Levelized Cost of Electricity (LCOE - the total cost of building and operating a power plant divided by the total energy it produces over its lifetime) told a stark story. Solar power had become

41 percent cheaper than the cheapest fossil fuel alternative. Wind power was 53 percent cheaper.[3] This is the definition of an economic discontinuity. When the new technology becomes cheaper than the marginal cost of operating the old technology, the transition accelerates dramatically. Utilities don't wait for old plants to reach the end of their planned lifetime. They shut them down early because every day they operate is money lost compared to alternatives.[16]

The implications compound. As old fossil fuel plants close, the industry loses scale. Revenues decline, making it harder to invest in maintaining remaining infrastructure. Skilled workers retire or retrain for renewable industries. Hiring dries up because no one wants to work for a dying industry. Supply chains atrophy. Political influence wanes. Lobbying money dries up. The death spiral accelerates. Meanwhile, renewables enter a virtuous, upward cycle. More deployment drives more manufacturing scale, which reduces costs further, which drives more deployment, by more and better trained construction crews. Each gigawatt installed makes the next gigawatt cheaper and easier to integrate.

The Storage Revolution

For decades, critics of renewable energy had a trump card: intermittency. The sun doesn't always shine. The wind doesn't always blow. How can you run a modern economy on unreliable power? It was a fair question, and for most of renewable energy's history, it didn't have a fully satisfactory answer. Grid operators could use natural gas "peaker" plants to fill gaps. They could build transmission lines to move power from where the sun was shining to where it wasn't. They could use pumped hydroelectric reservoirs as giant batteries. But these solutions had limits, reduced efficiencies and added cost.

Then something unexpected happened. The electric vehicle revolution created a massive manufacturing boom in lithium-ion batteries. Tesla, BYD, and dozens of other companies built gigafactories to meet the demand from millions of EVs. And batteries, like solar panels, follow Wright's Law. Global average turnkey energy storage system prices fell 40 percent from 2023 to 2024, reaching $165 per kilowatt-hour, marking the biggest single year drop since tracking began.[17] In just over a decade, battery costs declined roughly 85 percent.[18] Suddenly, grid-scale storage wasn't just a futuristic concept, it was also economically viable.

The intermittency problem that had plagued renewables for generations was solved not by a breakthrough in renewable technology, but by the automobile industry's manufacturing scale creating a breakthrough in storage technology pricing. It was an accidental revolution, but no less real for being unplanned. By 2025, utilities routinely paired new solar farms with battery storage systems. The combination could deliver power on demand, just like a fossil fuel plant, but at a lower cost and with zero emissions. The last major technical argument against renewables evaporated.

One more thought. Battery chemistry is not static. Tremendous amounts of time and money are being put into new battery chemicals that will make batteries that are lower in cost, more efficient and last longer. It will be like going from prop planes to jets. The entire cycle of Wright's Law is about to happen to new battery technologies who's "starting line" is often the finish line for older chemistries.

The Acceleration Ahead

Understanding Wright's Law reveals something crucial about the future: renewable energy costs will continue falling, and the pace of deployment will continue accelerating. Every doubling of cumulative solar capacity leads to another 20 percent cost reduction. We're now adding terawatts of solar globally. Each doubling happens faster than the last because the base is larger and manufacturing capacity is expanding. This means we're moving through multiple doublings per decade, each one making solar cheaper and more competitive.

The same applies to batteries, wind turbines, and the associated technologies of the renewable revolution. Each follows its own learning curve, and each curve accelerates the others. Cheaper batteries make solar more valuable because it can be stored. More renewable deployment creates demand for grid integration technologies, which themselves get cheaper with scale.

The Implications Are Staggering

What does this mathematical inevitability mean in practice? It means that within a decade, it will be difficult to find any circumstance where electrical generation by fossil fuels makes economic sense for new capacity. By 2030, we might see the first major economy where even maintaining existing fossil fuel plants becomes economically irrational compared to replacing them with renewables plus storage. It means that developing countries, rather than following the industrialized world's path through a fossil fuel phase, will leapfrog directly to renewable energy. Why import expensive, foreign fossil fuel when cheaper, domestically produced solar and wind energy are available? It means that the trillions of dollars in fossil fuel reserves still in the ground may never be extracted, not because of environmental regulation, but because they are economically worthless. The oil age won't end because we run out of oil. It will end because we won't pay for it.

A World Transformed by Mathematics

Chapter one described 2025 as the year everything changed. This chapter reveals why: the mathematics of manufacturing learning curves crossed the mathematics of resource extraction in a way that makes the renewable revolution unstoppable.

The next chapter will explore how this plays out in the real world of politics, policy, and power. Because while the math is clear, the human response to revolutionary change is never simple. Political systems built around fossil fuel economics don't surrender easily. Trillions of dollars in stranded assets create desperate resistance. Regional economies dependent on oil and coal employment face wrenching transitions. Banks have trillions in loans tied to fossil fuel reserves as collateral.

But first, it was essential to understand the mathematical foundation that makes all that resistance ultimately futile. Wright's Law isn't on anyone's side. It just is. And it has chosen renewables.

CHAPTER 3

The Market Takes Over from Politics

On a windswept plain in West Texas, thousands of white turbines spin silently against the blue sky. Each one generates enough electricity to power hundreds of homes. Together, they represent billions of dollars in investment. And they were built in one of the most politically conservative regions in America, in a state whose Republican leadership has deep ties to the oil and gas industry, during a period when renewable energy faced open hostility from the state legislature.

This is the paradox that reveals the most important truth about the renewable energy revolution: it no longer depends on politics. The massive wind farms of West Texas didn't arise because of progressive environmental policy. They arose despite political opposition, because the economics were overwhelming. When a rancher can lease land for turbines and make more money than from cattle, the turbines get built. When utilities can generate electricity at half the cost of natural gas, they build wind farms regardless of what legislators prefer. When investors see better returns in renewable projects than fossil fuel projects, capital flows toward renewables regardless of which party controls government.[11]

This chapter examines the single most important development in the renewable energy transition: the moment when market forces became strong enough to override political opposition. Once that threshold was crossed, the renewable revolution became unstoppable.

The Texas Case Study

Texas provides the clearest illustration of markets overpowering politics in energy. The state produces more oil and gas than many OPEC nations. Its political leadership is firmly Republican and generally skeptical of climate policy. The state legislature has repeatedly attempted to curtail renewable energy development with various bills designed to protect fossil fuel interests. And yet, despite this hostile environment, more capacity was added to the Texas grid in 2024 from solar and battery storage than any other source. The state set records for solar generation, wind production, and energy storage discharge in early 2025. Texas generates 7,016 percent more solar power and roughly 200 percent more wind power than a decade ago.[19]

How did this happen in an environment of political hostility? Economics. Pure economics. Texas has extraordinary wind resources in West Texas and the Panhandle, and excellent solar resources across much of the state. The state's deregulated electricity market allows generators to compete directly on price. There's no regulatory commission insisting on "all of the above" strategies or mandating baseload coal plants. Generating technology that can produce the cheapest electricity wins.

And renewable generators win because they produce the cheapest electricity. Wind farms in Texas regularly sign power purchase agreements at $20 to $40/MWh range.[20] Natural gas plants typically require $45 to $74/MWh to be profitable.[21] When the math is that stark, ideology becomes irrelevant. Even oil and gas companies are investing in Texas renewables. Not because they've gone green, but because that's where the returns are. When Occidental Petroleum and Chevron invest in solar farms, they are not making political statements, they are following the money.[22]

Corporate Procurement: The Trillion-Dollar Shift

Perhaps nothing demonstrates the market-driven nature of the renewable transition more clearly than corporate renewable energy procurement. In the past five years, corporations have become the largest drivers of new renewable energy development, and they are not waiting for governments to mandate it. Amazon, Microsoft, Google, and Meta have collectively committed to procuring more than 50 gigawatts of renewable energy. More than the total generation capacity of many countries. Apple runs entirely on renewable energy across its global operations. General Motors committed to 100 percent renewable energy by 2035.[23] Walmart is targeting 100 percent renewable energy by 2035.[24] Why? These companies haven't suddenly become environmental charities. They are making cold, calculated accounting decisions.

First, renewable energy has become the least expensive energy. Long-term power purchase agreements with solar and wind farms lock in low electricity costs for 20 to 25 years with no fuel price risk. Fossil fuel plants cannot offer that certainty. Natural gas prices fluctuate, coal availability changes, carbon regulations might arise. Renewables provide price stability.[25]

Second, corporate renewable commitments attract capital. ESG (Environmental, Social, and Governance) investing has become a massive force in global financial markets. Pension funds, sovereign wealth funds, and institutional investors increasingly screen for climate risk. Companies that can demonstrate climate leadership attract lower-cost capital.

Third, renewable energy enhances corporate reputation with minimal cost. Employees want to work for environmentally responsible companies. Customers increasingly factor environmental performance into purchasing decisions. Renewable energy procurement provides reputation benefits at lower cost than fossil fuel alternatives.

These choices are market calculations, not political positions. And they are accelerating. Corporate renewable procurement in 2024 reached 68 gigawatts globally, a new record.[26] That's 68 gigawatts of renewable capacity being built because corporations see it as the best business decision, regardless of government policy. This matters enormously for the speed and irreversibility of the transition. Governments change. Policies reverse. But corporate investments are long-term commitments based on economics. Once made, they create self-reinforcing momentum.

The Financial Sector's Climate Reckoning

Money sees the future before politics does. And in 2024 and 2025, financial markets spoke clearly: fossil fuels are riskier assets, renewables are safer, growth investments. Global investment in renewable energy and storage exceeded $2.4 trillion in 2024, roughly double the investment in fossil fuel production and infrastructure.[13] This wasn't primarily government spending; it was private capital making calculations about where returns would come from in the future. Major banks have begun declining to finance new coal projects, not because of environmental activism, but because of stranded asset risk. If coal plants become economically unviable before their planned lifetime ends, lenders don't get repaid. Why finance an asset that might be shut down early when cheaper alternatives exist?

Pension funds and university endowments started divesting from fossil fuels for purely fiduciary reasons. Returns from fossil fuel companies have lagged broader market returns as renewable competition intensified. Fund managers have an obligation to maximize returns for beneficiaries. When fossil fuel stocks underperform, divestment becomes a financial decision, not a political statement. This wasn't environmentalism, it was risk mitigation. Fossil fuel assets face growing risks from technological disruption, policy changes, and physical climate impacts. Smart money reduces exposure to high-risk assets.

By 2025, this trend created a feedback loop. Reduced investment in fossil fuels means less money for maintaining infrastructure, developing new fields, or improving technology. This accelerates the cost disadvantage of fossil fuels. Meanwhile, massive investment in renewables drives continued cost declines, making renewables even more competitive. The gap widens every year. Financial markets have essentially decided the energy transition is happening and are positioning accordingly. No government can override the collective judgment of global capital markets for long. When trillions of dollars flow one direction, that direction becomes reality.

When Economic Reality Overrides Political Preference

The renewable energy transition has reached the point where economic forces are strong enough to continue driving change even in hostile political environments. This is the breakthrough moment that makes the transition irreversible. Like a roller coaster just passing over the top of the first drop. Consider what this means in practice:

A utility executive in a red state might personally prefer coal power and vote for politicians who support coal. But when they evaluate options for new generating capacity, they choose solar because it is cheaper. Their job is to provide affordable electricity, and renewables do that better than coal. Personal preference becomes irrelevant.

A rancher in Texas might be skeptical of climate change and hostile to environmental regulation. But when a wind developer offers to lease land for turbines at payments that exceed livestock revenue, the rancher signs the lease. Economic interest overrides ideology.

A pension fund manager might have no personal interest in environmentalism. But when fossil fuel stocks underperform and renewable energy companies show stronger growth, fiduciary duty requires shifting allocations. Obligation to beneficiaries overrides personal views.

Oil company executives might have spent their career in fossil fuels and would prefer to keep doing so. But when investors demand climate risk disclosure and banks become reluctant to finance new fossil fuel projects, the company diversifies into renewables to access capital. Market pressure overrides company legacy.

These individual decisions, multiplied across millions of economic factors, create an overwhelming force for change that operates independently of political systems. Governments can slow this process with subsidies for fossil fuels or barriers to renewable development. But they cannot stop it without massive, overt and sustained intervention that most political systems won't support in the long term.

Even authoritarian governments with strong state control of energy systems, like China and the Gulf states, are investing heavily in renewables. They can see the economic direction and are positioning to maintain competitive industries. Saudi Arabia, with the world's lowest-cost oil production, is investing tens of billions in solar energy because they recognize economic reality.[27]

The Stranded Asset Problem

For fossil fuel interests, the market transition creates what economists call a stranded asset problem, and it is getting worse every year. Stranded assets are investments that lose value prematurely due to changes in market conditions, technology, or regulation.

A coal plant built in 2015 with an expected 40-year operating life might have seemed like a sound investment at the time. But by 2025, that same plant can be uneconomical compared to new solar farms. Utilities face a choice: continue operating an asset that loses money every day or shut it down early and write off remaining value. The $1.3B Comanche 3 power plant in Pueblo, Colorado was built in 2010 with an expected life of sixty years. It is now slated for closure in 2031. In all fairness, this is not just an economic decision as the plant has been unreliable and has been shut down for periods of time, but it illustrates the economic risk utilities and financiers take in investing in fossil fuel projects today.

The scale of potential stranded assets in fossil fuels globally is staggering. Estimates range from $1 trillion to $4 trillion depending on how quickly the transition proceeds. This represents enormous financial risk for anyone holding those assets.[28] Smart investors exit before taking losses. This selling pressure further depresses fossil fuel asset values and makes new investment even less attractive.

For renewable energy investors, the dynamic is opposite. As renewable costs fall and deployment accelerates, existing renewable assets become more valuable because the grid increasingly needs what they produce. Early renewable projects that seemed risky now look prescient. This creates positive momentum for continued renewable investment.

The Irreversible Threshold

Markets have reached what might be called the irreversible threshold; The point where economic forces driving the renewable transition are strong enough to continue even if political opposition intensifies. The roller coaster moment. This doesn't mean politics is irrelevant. Government policy absolutely affects the speed of transition. Supportive policies like tax credits, streamlined permitting, transmission investment, and carbon pricing accelerate change. Hostile policies like fossil fuel subsidies, slowed or denied approvals, renewable energy barriers, and transmission restrictions slow change. In Australia home solar connection applications are approved in a week, in the US it can take six months.

But slow does not mean stop. The underlying economic forces (Wright's Law cost declines, stranded asset dynamics, corporate procurement, and financial sector reallocation) continue regardless of political environment. A renewable transition that might take 20 years with supportive policy might take 30 years with hostile policy, but it happens either way because the math is overwhelming. This is the fundamental difference between the current renewable transition and previous "green" waves. Previous attempts depended primarily on policy support and could be reversed when political winds shifted. The current transition is driven primarily by economics with policy as an accelerant or brake, not the primary motive force.

Looking Forward

Part II of this book will examine the global landscape of renewable deployment. How different regions and countries are navigating this market-driven transition given their unique circumstances, resources, and political systems.

But understanding that markets have taken over is essential context for everything that follows. When we look at China's massive solar manufacturing dominance, it is not primarily a story about government industrial policy (though that plays a role). It is a story about recognizing opportunity and capturing market share in the fastest growing industry in the world.

When we examine Europe's renewable integration challenges, it is not primarily about climate policy (though that matters). It is about how to maintain industrial competitiveness while managing a rapid transition driven by both policy and economics.

When we analyze developing countries leapfrogging to renewable energy, it is not about environmental virtue. It is about choosing the cheapest path to electricity access.

The financial markets have spoken. The renewable transition is happening because it makes economic sense, not because it is morally right or environmentally necessary (though it is both). Those additional justifications are bonuses. The primary driver is the oldest force in economics: rational decision makers choosing less expensive options over more expensive ones.

THE GLOBAL
RENEWABLE LANDSCAPE

CHAPTER 4

China Becomes
The Renewable Superpower

In the desert regions of Xinjiang province, solar panels cover the dry landscape. In the windswept grasslands of Inner Mongolia, wind turbines dot the landscape like steel forests. In the manufacturing hubs of Jiangsu and Anhui, gigafactories operate around the clock, producing solar panels at a scale that would have seemed impossible a decade ago. China has become the undisputed superpower of renewable energy and understanding how it got there (and what it means for the rest of the world) is essential to understanding the global energy transition.

Most people associate China with manufacturing prowess, but China's dominance in renewable energy is more than manufacturing. It is about a comprehensive, top-to-bottom, government driven, industrial strategy that recognized the renewable revolution early, invested massively, and created self-reinforcing advantages that will shape global energy markets for decades. Historically, China had made things that others invented; this time China is both the innovator and the manufacturer.

In describing China's renewable energy emergence, all the metrics are superlatives. Today China produces 80% of the world's solar PV equipment and is home to 58% of the world's installed solar capacity.[29] China's additions to solar capacity in 2024 were more than twice the entire amount of utility solar installed in the US to date.[30]

In one month, April, 2025, China installed more solar capacity than Australia had in its history.[31]

But China's renewable energy story isn't just about scale. It is about how a nation transformed from a renewable energy follower to the global leader in less than two decades, and what that transformation reveals about the economics and geopolitics of the energy transition.

From Follower to Leader: The Strategic Shift

Twenty years ago, China was a climate bad guy in Western eyes. It was the world's largest coal consumer, rapidly increasing emissions and lead by a government that was seemingly unconcerned about environmental consequences as smog choked its cities. It was Western nations that led in renewable technology development. Germany pioneered large-scale solar deployment. Denmark mastered wind energy. American and European companies dominated manufacturing.

Then behind closed doors, something changed. At some point, Chinese policymakers recognized that renewable energy wasn't just an environmental issue, it was an economic and geopolitical opportunity. China had to import oil and natural gas that often came from politically unstable regions. Renewables, on the other hand, could be manufactured domestically and the fuel (sun and wind) were free. But more importantly, there was a realization that the country that dominates renewable manufacturing going forward would control the energy technology of the future, the same way that nations that had controlled oil production shaped 20th century geopolitics. With renewable energy, China wanted to become with renewable energy, what Saudi Arabia has been with oil.

China is becoming the world's clean energy superpower; the first electrostate (as opposed to existing petrostates).[31] This wasn't environmentalism (though reducing China's catastrophic air pollution provided some motivation). This was ambitious and effective centralized industrial policy. It was well suited to China's hybrid governmental approach that featured centralized planning with Darwanistic execution.

The strategy had several components. First, massive domestic subsidies and mandates created a guaranteed market for renewable installations. This gave Chinese manufacturers an immediate scale advantage, where they could operate at volumes that drove costs down faster than competitors. Second, government-backed financing made capital cheap for renewable manufacturers, allowing rapid expansion even when profit margins were thin. Third, strategic investment in the entire supply chain. From polysilicon production to module assembly, an integrated manufacturing ecosystem was created that foreign competitors would be hard pressed to replicate in short order. Fourth, and perhaps most importantly, China allowed brutal market competition among domestic manufacturers. Unlike industrial policies that protect national champions from competition, China let hundreds of solar companies compete fiercely, driving innovation and efficiency improvements. Many failed. The survivors were lean, efficient, and globally competitive. Darwin would be proud.

The result? By 2025, China holds over 80% of global solar manufacturing capacity, producing end product so cheaply that competitors in other countries struggle to survive even with governmental subsidies and tariff protection. Chinese solar panels regularly cost 40-50 percent less than panels manufactured in the United States or Europe.[32] It is hard to imagine a scenario where China will lose this leadership position.

The Manufacturing Juggernaut

A walk through a modern Chinese solar panel factory will let you understand why they are dominating this industry. These facilities operate at scales unimaginable in Western manufacturing. A single factory might produce 20 to 30 gigawatts of panel capacity annually. More than the entire solar installation totals of most countries. Automation is extensive, with robots handling most assembly steps.

The supply chain is vertical and local with silicon refiners, wafer producers, cell manufacturers, and module assemblers often located in the same province, sometimes the same industrial park.

This geographical concentration creates advantages that go beyond simple logistics. Engineers from different companies share insights (often by changing employers). Manufacturers of production equipment can rapidly offer improved designs based on factory feedback and government officials can coordinate infrastructure investments (power, water, transport) to support the entire ecosystem. These manufacturing advantages extend beyond solar panels. China produces most the world's wind turbines, batteries, inverters, and other renewable energy equipment. In batteries particularly, Chinese companies like CATL and BYD have massive scale advantages. CATL alone produces more battery capacity than all Western manufacturers combined. Tesla is CATL's largest customer.

This isn't just about cheap labor (much of the manufacturing is highly automated). It is about scale, supply chain integration, manufacturing expertise, and yes, strategic subsidies. Western governments often complain about unfair Chinese subsidies. They are not wrong, because China does subsidize strategically, as seed money. But they miss the larger point. At the scale Chinese manufacturers now operate, subsidies matter less than they once did. The pump has been primed, and Wright's Law has kicked in. Chinese manufacturers are moving down the cost learning curve faster than competitors, and the advantage compounds every year. Western government, mired in internal politics, are letting the Chinese run away with the future.

Domestic Deployment: Building the World's Largest Renewable Grid

China's manufacturing dominance would matter less if it only exported equipment. But China is also deploying renewables domestically at unprecedented scale. According to China's National Energy Administration, the country's photovoltaic capacity increased by 45.7 gigawatts in the first quarter of 2024, taking the country's total photovoltaic capacity to 660 gigawatts, while wind power capacity was around 460 gigawatts. That's over 1,100 gigawatts of wind and solar capacity.[33] More than the United States, European Union, and India combined. And China is adding capacity faster than anywhere else on Earth. Why the massive deployment? Several reasons converge:

First, China's air pollution crisis became politically untenable. Smog in Beijing and other major cities wasn't just an environmental problem, it was causing public health disasters and social unrest. Coal-fired power plants were major contributors and an easy target for criticism. Renewables offered a path to cleaner air.

Second, energy security. China imports most of its oil and much of its natural gas. These imports flow through vulnerable shipping lanes (the Straits of Malacca are a potential chokepoint in any geopolitical conflict). Renewables manufactured domestically from abundant Chinese minerals eliminate this vulnerability. The payoff was seen during 2025's "Golden Week" holiday. Usually, automobile travel associated with Golden Week triggers a huge upswing in gasoline usage, but in 2025, there was actually a drop as a significant percent of the Chinese population now travels by hybrid or EV. And it didn't hurt that China had added 18 million new charging ports in the previous twelve months.

Third, and most importantly, economics. As Chinese manufacturers drove down costs through scale and learning curves, renewables became the cheapest option for new generating capacity. Chinese utilities now build solar and wind because they are cheaper than coal or gas, not because the government mandates it (though mandates helped establish the market initially). Renewable and green energy now accounts for about 10% of China's economy, a number greater than the value of the real estate market in China and low electricity costs have created a structural cost advantage for all Chinese manufacturing concerns.

China's grid operators are learning at scale how to manage high-renewable penetration. They are developing the world's most sophisticated grid management systems because managing 660 gigawatts of solar requires capabilities that didn't exist a decade ago. This expertise itself becomes an export commodity. Chinese companies will help other nations integrate renewables because they've solved problems at scale that others are just encountering. But more importantly, China is creating an industrial expense advantage with low-cost electricity, which when added to its existing advantages, could make China unstoppable economically.

The Geopolitical Implications

China's renewable dominance creates fascinating geopolitical dynamics. On one hand, it is good for global climate goals and the sustainability of human life on the planet. Cheap Chinese equipment makes renewable deployment affordable everywhere, accelerating the energy transition. Low-cost energy would make an attractive addition to China's existing Belt and Road Initiative and create economic dependencies that make Western governments nervous.

When 80 percent of solar panels come from one country, that country has leverage. If geopolitical tensions escalate (e.g. China invades Taiwan), could China restrict exports to limit international dissent? Or use renewable equipment sales to bring diplomatic pressure? The concerns aren't hypothetical. China has used rare earth mineral export restrictions for political purposes before. These factors have sparked efforts to build alternative supply chains. The United States passed the Inflation Reduction Act partly to subsidize domestic renewable manufacturing. Europe launched similar initiatives. India is trying to build its own solar manufacturing capability and these efforts will succeed at some level. But they face the fundamental challenge of competing with Chinese manufacturers who are years ahead on the learning curve and operating at vastly larger scale.

A more likely outcome is one where regional supply chains will serve domestic markets with higher-cost equipment, while Chinese manufacturers dominate the cost-sensitive global market (particularly in developing countries that prioritize affordability over supply chain security). This creates a split global market with Chinese equipment priced 30-50 percent lower than alternatives, and countries choosing between cheap Chinese gear with geopolitical risk, or expensive domestic gear with supply chain security. Like arms sales today. For climate goals, this is a desirable outcome as both paths lead to renewable deployment. For geopolitics, however, it is complex. China gains influence but also has incentives to keep prices low and supply reliable because if they are seen as unreliable suppliers, alternatives accelerate.

Peak Emissions: Ahead of Schedule

Perhaps the most significant implication of China's renewable buildout is that the world's largest carbon emitter appears to be peaking emissions years ahead of international commitments. China pledged to peak emissions by 2030. The data suggests it happened in 2024-2025. This is almost unprecedented. China's rapid economic development was supposed to drive emissions growth for years to come, but instead, renewable deployment is outpacing electricity demand growth, meaning additional power needs are being met with clean energy while coal generation gradually declines. Why ahead of schedule? As we have discussed, not climate action, but economics. Chinese utilities prefer to build solar farms over coal plants because solar is cheaper, cleaner (addressing the air pollution crisis), and faster to construct. Coal plants take years to permit and build. Solar farms can be deployed in months.

China's progress has global implications. If China (still a developing economy, still heavily invested in coal infrastructure) can peak emissions through renewable deployment driven primarily by economics, other countries could as well. China, the former climate bad guy, can become the poster boy for renewable transition that works at massive scale, in a developing environment while supporting rapid economic growth.

The lessons learned in China about grid integration, storage deployment, and managing high renewable penetration are being exported globally through Chinese companies building renewable projects worldwide. China has become not just the manufacturer but also the practitioner, proving renewable dominance is achievable at any scale.

The First "Electrostate"?

China's trajectory points toward becoming what some analysts call an "electrostate", a 21st century counterpart to 20th century petrostates. Where petrostates derived geopolitical power from controlling oil production, electrostates will derive power from controlling renewable energy manufacturing, battery production, and the minerals required for clean energy technology, as well as the expertise in technology integration.

China is positioning itself as the first electrostate by dominating every link in the renewable energy supply chain:

- Minerals: China controls most of the rare earth mineral production and processing. It is securing lithium, cobalt, and other battery minerals through investments in mines worldwide, particularly in Africa and Latin America.
- Manufacturing: Dominant in solar panels, wind turbines, batteries, inverters, and other renewable equipment. This manufacturing base gives China the same kind of monopolistic economic leverage that OPEC once had through oil, specifically, the ability to control global prices and supply.
- Technology: Investing heavily in next-generation technologies like advanced batteries, green hydrogen production, and grid management systems. Chinese companies file the most patents in these areas globally.[34]
- Deployment expertise: Operating the world's largest renewable grid gives Chinese companies unmatched practical experience, making them preferred partners for renewable projects worldwide.

This isn't just about current market share. It is about controlling the technologies that will define the coming Renewable Century. The petrostates of the 20th century wielded enormous geopolitical influence because industrial economies depended on their oil. China is building similar dependencies in renewable energy technology.

But there is also a difference. Renewable energy is more distributed and available than fossil fuels. While oil comes from specific geographic regions, sun and wind are everywhere. So, China's electrostate power is different. It is not about controlling resources, but about controlling manufacturing, technology, and integration expertise. Other countries can deploy renewables, but they'll depend on Chinese equipment, batteries, and increasingly, Chinese financing and technical prowess. Petrostates sold a commodity product to end users. China's electrostate is more of a franchise business that provides equipment for other countries to become mini-electrostates. Like drug dealers, petrostates created dependency. Electrostates, on the other hand, will create energy independence and self-determination. It is certainly ironic that the United States, the historic bastion of democracy is retrenching on fossil fuels and addictive trade relationships, while China is offering energy independence. This should make China's renewable dominance more resilient than petrostate power. Oil exporters constantly worry about reduced demand or reserves running low, but renewable equipment manufacturing gets more competitive with scale. China's advantages compound over time rather than depleting like oil reserves.

What This Means for the Global Transition

China's renewable superpower status accelerates the global energy transition. By driving down costs through massive scale manufacturing, China makes renewable deployment affordable in developing countries that couldn't otherwise afford it. A solar panel that costs $0.15 per watt in China versus $0.25 per watt from Western manufacturers might not matter much to wealthy nations, but for Kenya, Bangladesh, or

Nigeria, that price difference determines whether they can afford solar at all. China's willingness to export not just equipment but also financing (through the Belt and Road Initiative and other programs) means developing countries can deploy renewables without large capital outlays. This speeds electrification globally, bringing power to hundreds of millions without building fossil fuel infrastructure. The unspoken price is political allegiance to China.

For climate goals, China's path is crucial. If the world's largest emitter and biggest developing economy can rapidly deploy renewables and reach peak emissions, it demonstrates feasibility for every other nation. China becomes the proof of concept that economic development and emissions reduction are compatible, not through sacrifice, but through economics. And they are self-reinforcing. The geopolitics are complex, but the climate math is simple: China's renewable dominance is driving costs down and deployment up, globally. That accelerates the transition away from fossil fuels faster than would happen with more distributed manufacturing. It creates dependencies, but they are different dependencies than fossil fuel reliance. Less zero-sum, more widely distributed, based on manufacturing rather than geology. In truth, China is showing laser focused leadership while the legacy Western democracies have fist fights over cultural issues.

What's Ahead

China's renewable superpower status will shape global energy markets for decades. The country that controlled 20th century energy was Saudi Arabia and they sold oil to oil addicted countries. The country controlling 21st century energy will almost certainly be China. But the nature of that control is different; less about owning scarce resources, more about manufacturing at scale and accumulating technological expertise. More like a franchise, less like a drug dealer.

As we'll see in the next chapter, Europe is taking a different path, less focused on manufacturing dominance, more focused on creating the grid systems and market structures to integrate extremely high renewable penetration. The European model proves that renewable-dominated grids can work in modern industrial economies. The Chinese model proves that massive scale manufacturing can drive costs down to the point where renewables become economically dominant globally.

Together, these two approaches, European integration excellence and Chinese manufacturing dominance, are driving progress and making the renewable transition irreversible at global scale.

China didn't become a renewable superpower by accident. It was a national strategy, executed at scale, over two decades. The result is an electrostate that will shape global energy as profoundly as OPEC shaped 20th century oil markets. But this time, the power comes from making things rather than extracting things. And that makes all the difference.

CHAPTER 5

Europe Proves the Transition Works

On a windy afternoon in March 2024, something remarkable happened in Denmark. For several hours, wind turbines generated more electricity than the entire country could consume. Grid operators exported the excess to Germany and Sweden, profiting off excess electricity. A decade earlier, this event would have caused grid instability and required shutting down turbines. Today, sophisticated grid management, international interconnections, and smart demand response made the high-wind period an opportunity rather than a problem.

Electricity generation from renewable sources reached a record 46.9% in the European Union in 2024, with Denmark leading at 88.4% renewable penetration, followed by Portugal at 87.5%.[35] These aren't small, isolated grids with modest electricity demands. These are highly industrialized nations reliably running sophisticated economies on renewable-dominated power systems. While China dominates renewable manufacturing, Europe dominates on grid integration. Demonstrating to the rest of the world how to actually run a power system when renewables provide most of the electricity.

Europe's path to renewable dominance took a different route than China's. It wasn't about building manufacturing capacity or achieving geopolitical leverage. It was about climate leadership, sustainability, energy independence (particularly after Russia's invasion of Ukraine), and codifying the regulatory and technical frameworks to make high-renewable grids work reliably.

The European Model: Policy-Led, Market-Driven

Europe pioneered the policy tools that made renewable deployment economically viable, is now letting market forces take over once renewables became cost competitive. Feed-in tariffs in Germany and Denmark in the 1990s guaranteed prices for renewable electricity, providing certainty that attracted investment. National renewable portfolio standards required utilities to source specific percentages of electricity from green energy. Carbon pricing made fossil fuels more expensive, leveling the playing field and priming the pump. Europe became the world's first incubator for renewable energy.

These policies worked, but with rough spots along the way. Germany's Energiewende (energy transition) drove massive solar deployment but also created market imbalances. Wholesale electricity prices dropped as solar flooded the grid on sunny days, making conventional plants unprofitable but still necessary for backup. Grid congestion emerged as wind-rich northern regions couldn't transmit power south to industrial centers. Consumer electricity prices rose as the costs of grid upgrades and renewable subsidies flowed through to bills. Critics pointed to these challenges as proof that renewable transitions were impractical, but these kinds of hiccups are common with new technology. The early aviation industry was a mess for decades, but the airplanes we use today are indispensable and practical. Modern roads and the interstate freeway system took decades to roll out. Renewable energy critics miss the larger story. Germany's messy start created an opportunity to solve problems that every grid would eventually face as renewables scaled. The experience was gained, the technology developed and the market structures were created that became templates for the world. Darwinism applies to technology, as well as animals.

By 2025, many of Germany's early challenges are either solved or greatly mitigated. Grid congestion is being addressed through massive transmission investments and better demand management. Subsidy costs are falling as renewables become cheaper than fossil fuels without pricing support. Grid stability with high renewable penetration is no longer theoretical, it is demonstrated reality across multiple European countries. Wind and solar overtook fossil fuels in EU electricity generation for the first time in the first half of 2024, reaching 30% of generation while fossil fuels fell below that level.[36] The crossover happened, and renewables now provide more electricity than fossil fuels in the world's second-largest economy. The lights stayed on. Industries kept running. Economic growth continued.

Denmark: The 88% Solution

Denmark provides the most dramatic proof of concept. With 88.4% of electricity from renewables in 2024, mostly wind, Denmark demonstrates that a modern industrial nation can run almost entirely on renewables while delivering one of the most reliable grids in the world.

How has this happened? Several factors combine:

- Strong interconnections: Denmark's grid is tightly integrated with the neighboring countries of Norway, Sweden, Germany. When Danish wind generation exceeds demand, excess electricity exports. When wind is low, Denmark imports. These interconnections effectively create a larger grid that smooths out local variations.
- Flexible demand: Danish industrial users have real-time electricity pricing contracts that incentivize them to use more power when it is cheap (windy) and less when it is expensive (calm). This demand flexibility effectively provides grid stability without large batteries or backup plants.
- Modern grid management: Danish grid operators use AI-powered forecasting to predict wind generation hours in advance, allowing them to position the grid optimally. When forecast accuracy reaches 95%+ for hours-ahead predictions, intermittency becomes manageable.
- Geographic advantage: Denmark is small, making grid management easier than in large countries. It is surrounded by

countries with different generation profiles (Norway's hydro, Germany's "all of the above", Sweden's nuclear and hydro), making interconnection valuable.

But what matters most is that Denmark didn't sacrifice economic competitiveness. It maintains a modern industrial economy, exports wind turbines globally (Vestas and Siemens Gamesa are partially Danish), and has demonstrated to the world that 88% renewable electricity is practically achievable.

Portugal: The Renewable Diversity Approach

While Denmark achieved high renewable penetration mostly through wind, Portugal demonstrates a more diverse, "all of the above" approach. Portugal reached 87.5% renewables in 2024, combining wind, solar, and hydro. This diversity matters because it shows that countries don't need to rely on one dominant resource. Multiple renewable sources can complement each other and almost always results in a more robust grid.

Portugal's renewable success stems from geography and good policy. The country has excellent wind resources on its Atlantic coast, strong solar potential in the south, and existing hydroelectric dams that provide dispatchable power and storage. Portuguese policymakers recognized these complementary resources and created auction systems that drove deployment, and financial health across all three technologies. The lesson was renewable transitions don't require perfect resources in every category. Portugal isn't the windiest country or the sunniest. But it has enough of both, plus existing hydro, to reach nearly 90% renewable electricity. This suggests that most countries have sufficient renewable resources if they deploy diversely rather than focusing on a single technology. Diversity of resources is a strength.

Portugal also demonstrates another crucial point; high renewable penetration didn't kill industrial competitiveness.

Portugal's economy grew even as it transitioned away from fossil fuel generation. Manufacturing continued. Energy-intensive industries adapted. The apocalyptic warnings that renewable transitions would deindustrialize Europe proved false; at least for countries that managed the transition competently.

Germany: Learning Expensive Lessons

Germany's Energiewende is the most analyzed, most expensive, most challenging, and ultimately most instructive renewable transition story. Germany set ambitious goals (80% renewable electricity by 2030, 100% by 2050), backed them with massive policy support, and encountered virtually every problem that can arise in renewable transitions.

Early German solar subsidies were too generous, leading to over-deployment and enormous costs. Grid infrastructure lagged deployment, creating bottlenecks. The simultaneous decision to phase out nuclear power (post-2011 Fukushima) removed clean baseload generation, forcing more reliance on coal and gas during the transition. Russian natural gas dependence created vulnerability that came back to haunt Germany after Russia's invasion of Ukraine. Prematurely removing perfectly good nuclear power plants was a mistake made by Germany and Japan which resulted in much greater carbon emissions from fossil fuel powered thermal power plants.

Critics love pointing to Germany's challenges. But they are missing the forest for the trees. Yes, Germany made mistakes. Yes, electricity prices rose (though they've come back down as renewables became cheaper). Yes, the transition took longer and cost more than projected. But Germany is succeeding with renewable electricity exceeding 62% in 2024, plummeting coal generation and a country that is on track to meet its 2030 renewable energy goals.[37]

The European Grid: Interconnection Enables Transition

Perhaps Europe's greatest contribution to proving renewable transitions work is the creation of a genuinely interconnected continental grid. This interconnection is crucial for high renewable penetration because it dramatically expands the geographic area over which generation and demand balance. When wind is strong in the North Sea (benefiting Denmark, Germany, Netherlands, Belgium), it is often calm in Spain and Italy. When the sun shines in Mediterranean regions, it may be cloudy in northern Europe. Interconnections allow this geographic diversity to smooth out variations in renewable generation.

The interconnected European grid effectively creates a continent-scale battery. Norway and Switzerland have massive hydroelectric storage capacity. When solar and wind generation across Europe exceeds demand, the excess can pump water uphill in Norwegian and Swiss facilities, storing it as potential energy. When renewable generation drops, these facilities release water through turbines, generating electricity for the continent. This cross-border cooperation required both technical infrastructure (high-voltage transmission lines, including undersea cables) and market infrastructure (coordinated electricity markets, cross-border trading mechanisms, shared grid management protocols). Creating this took decades and enormous investment. But it is paying off as high renewable penetration across Europe has been achieved partly because the "grid" is continental, not national. Much remains to be done, and the grid interconnection is a work in progress.

The lesson for other regions: renewable transitions are easier with large, interconnected grids (looking at you, Texas). North America has potential for similar continental-scale interconnection, as do regions of Asia and Latin America. Countries pursuing renewable transitions in isolation face harder challenges than those who coordinate with neighbors.

The Coal Exit: Accelerating Despite Challenges

Coal generation halved from 2016 to 2023 in the EU as wind and solar generation basically doubled. Coal's structural decline in Europe continues as 20% of the remaining EU's coal fleet was shut down in 2024 and 2025.[38] This coal exit is happening faster than predicted, driven primarily by economics. Renewable generation is cheaper, so utilities retire coal plants ahead of schedule. The energy crisis following Russia's invasion of Ukraine briefly slowed coal retirements as Europe sought to reduce Russian gas dependence, but the fundamental economic trajectory remained unchanged.

Germany's Ruhr Valley (for a century the heart of European coal and steel) is now increasingly a center for renewable energy technology companies and advanced manufacturing. The transition took decades and required significant public investment but demonstrates that coal-dependent regions can successfully transition if given resources and time. The renewable transition works best when there is a long-term commitment from political leadership.

Industrial Competitiveness: The Real Test

The most important question about Europe's renewable transition is whether it has maintained industrial competitiveness. If high renewable penetration made European industry uncompetitive when compared with rivals using cheaper fossil fuels, the European model would be a cautionary tale rather than inspiration. The evidence is still mixed in the frothy inflection point, but trending positive. European electricity prices spiked during the 2022 energy crisis, causing genuine competitiveness concerns and some industrial production shifting to regions with cheaper energy. But as renewable deployment accelerated and the crisis eased, prices normalized. By 2024, industrial electricity prices in Europe were declining again, and renewable generation was providing price stability that fossil fuel-dependent regions lacked.

Importantly, European industry is mostly adapting to high-renewable electricity. Energy-intensive industries like aluminum smelting, chemical production, and steel manufacturing are increasingly signing long-term power purchase agreements directly with renewable generators, locking in low prices and supply certainty. Some are locating facilities in high-renewable regions to benefit from cheap electricity during high-generation periods. Literally making hay while the sun shines.

European success stories show that industrial competitiveness with high renewable electricity is possible, but it requires industrial flexibility, smart contracting, and grid infrastructure that allows electricity-intensive production to locate where renewable generation is strongest.

Perhaps most telling is the fact that European companies have not abandoned Europe en masse for cheaper-energy regions. While some energy-intensive production has shifted (much to China, for cheaper labor and electricity), most European industry remains competitive. The apocalyptic predictions of deindustrialization due to renewable transitions haven't materialized. Europe remains the world's second-largest economy, maintains advanced manufacturing capabilities, and continues to export sophisticated industrial products globally.

What Europe Has Proven

By 2025, Europe has proven several crucial points:
- Very high renewable electricity is technically feasible in modern industrial economies. Denmark at 88%, Portugal at 87%, and EU-wide at 47% demonstrate this beyond doubt.
- Interconnection dramatically eases renewable integration. Countries connected to neighbors with different generation profiles and storage capabilities manage high renewable penetration much more easily than isolated grids.
- Industrial competitiveness can be maintained with high renewable electricity, though it requires adaptation and smart policy.

- Regulatory frameworks matter enormously. Good market design, grid codes, and cross-border mechanisms are as important as renewable generation capacity itself.
- Political persistence pays off. Despite challenges and setbacks, European commitment to renewable transitions persisted long enough for economics to take over from policy as the primary driver.
- Coal exit is economically inevitable once renewables reach sufficient scale, regardless of political preferences for coal.
- Expensive early lessons become cheap later knowledge. Germany's costly mistakes created expertise that benefits everyone pursuing renewable transitions.

Europe didn't achieve renewable dominance cheaply or easily. The transition has been expensive, challenging, and slower than hoped. But it is working. The lights stayed on. Industry competes. Emissions are falling dramatically. And the model is exportable as other regions can learn from European experiences without repeating European mistakes.

Looking Forward

Europe's renewable transition is approaching a new phase. With renewables already dominant in electricity, attention shifts to harder challenges: industrial decarbonization (hydrogen for steel and chemical production), heating (heat pumps replacing gas boilers), and ensuring that very high renewable penetration (70-80% or more) can work reliably year-round, not just on good renewable days. In November of 2025, the EU adopted a plan to hit 90% improvement in emissions by 2040, using 1990 as the baseline. This is a seismic step in the global transition to renewables and a strong pushback to fossil fuel "victory laps" in the US and some other countries.[39]

The next chapter will examine how the United States is navigating its own renewable transition. We will discuss the fragmented, market-driven, politically polarized path that nonetheless is producing dramatic renewable growth, particularly in unexpected places. The American approach is messier than Europe's coordinated strategy, but in some ways, just as effective. But Europe has proven the crucial point: renewable-dominated grids work in modern industrial economies. The question is no longer "can it be done?" but "how fast can everyone else do it too?"

CHAPTER 6

The United States: Fragmented but Advancing

The United States has not managed the energy transition the way Europe or China did. There's no coordinated national strategy, no continent-wide grid integration plan and no carbon pricing mechanisms that apply nationally. Instead, America's renewable revolution is happening through a chaotic, fragmented, sometimes contradictory mix of state-level policies, market forces, corporate decisions, and technological disruption. It is messy. It is inefficient. And somehow, it stumbles forward. Often in the least likely places.

This chapter explores how the United States is navigating its renewable transition despite political polarization, regulatory fragmentation, and a federal government that swings wildly between supporting and opposing climate action depending on which party controls the White House. The American story proves that market economics can drive renewable transitions even when the politics are hostile and coordination is minimal.

State Leadership: The Laboratories of Democracy

When the federal government couldn't or wouldn't lead on renewable energy, states have stepped in. The result is extreme variation. Some states are renewable leaders; others lag decades behind. But the leaders are proving what's possible and creating momentum that even resistant states cannot fully ignore.

California remains the prototype for progressive climate policy. In 2023, the state generated about 67% of its in-state electrical needs with solar, hydropower, and wind as the three largest sources of renewable power. California had 100 days in 2024 with 100% carbon-free, renewable electricity for at least part of each day.[40] The state's aggressive policies, including renewable portfolio standards, rooftop solar mandates for new construction and vehicle electrification requirements have created a massive market that drove costs down and proved that large economies can run primarily on renewables. Keep in mind that California's economy is the fifth largest in the world. Just behind Japan. Their advances with renewable energy are not trivial.

But California also demonstrates the challenges. Grid management with high solar penetration creates the "duck curve". The famous graph showing how net electricity demand drops midday when solar floods the grid, then ramps up steeply at sunset when solar disappears but people come home and turn on appliances. Managing this requires battery storage, flexible demand, and sophisticated grid operations. There have been real problems with supply not meeting demand in these transition times. Tweaking the grid, like Germany, will take a few years.

Texas, as discussed in Chapter 3, is the unlikely renewable powerhouse. Despite hostile state-level politics, Texas leads the nation in wind generation and is rapidly adding solar. The state's deregulated electricity market created conditions where cheap renewables thrive because they are simply the most economical option. Texas proves that market forces can drive renewable transitions even when government is actively trying to slow them down. In 2024, 92% of new electrical generation capacity came from renewables.[11]

Iowa gets over 64% of its electricity from wind, the highest percentage of any U.S. state.[41] This happened not through progressive climate policy (Iowa's state government is solidly conservative) but through economics. Iowa has excellent wind resources, available land, and supportive landowners who earn steady lease payments from turbines. Wind energy has revitalized rural economies, providing income to farmers facing volatile crop prices. The lesson learned is that renewable transitions can benefit rural conservative regions as much as urban progressive ones.

Florida is experiencing a massive solar boom despite having no renewable energy mandates and a state government generally hostile to climate policy.[42] Why? Economics and hurricanes. Solar-plus-battery systems have proved resilient when hurricanes knock out the central grid (an increasingly common occurrence). Homeowners and businesses install solar not for environmental reasons but for reliability and grid independence. And utility-scale solar farms are getting built because Florida's excellent sun and high electricity demand make solar profitable without subsidies.

America's state-level variations create a laboratory that demonstrates which policies work and which are failures. States that streamlined permitting expedited deployment. States that made interconnection easy attracted more distributed solar. States that invested in transmission got more utility-scale renewable development. This means other states can observe and copy the policies that worked while avoiding those that didn't.

Federal Policy: Swinging Wildly

Federal renewable energy policy in America swings dramatically with presidential administrations, creating uncertainty that normally would cripple long-term infrastructure investment. Yet somehow renewable deployment continues growing through these policy swings, demonstrating the underlying strength of market economics.

The Obama administration promoted renewables through tax credits, efficiency standards, and programs like the "cash for clunkers" stimulus that favored clean energy. The first Trump administration reversed directions by propping up coal and rolling back regulations favoring clean energy. The Biden administration swung hard back to renewables by passing the Inflation Reduction Act, the largest climate investment in American history. The IRA extended and expanded renewable energy tax credits, electric vehicle subsidies, and clean energy manufacturing incentives. And now, the second Trump administration put anything to do with renewables in their crosshairs for the harshest treatment possible. Even cancelling already approved wind turbine farm approvals retroactively, after the project was sustainably complete.[43]

Each swing was affected the pace of deployment but not the fundamental direction. Renewable capacity grew during the first Trump administration despite hostile federal policy because state-level policies and market economics continued driving deployment. The IRA accelerated growth significantly, but it boosted an existing trend as opposed to creating a new one. While the second Trump administration has done everything in their power to slow down renewables, the movement continues, albeit slowed down a bit.

This zig-zag path of American renewables demonstrates an important principle. Once renewable transitions reach economic viability, they become relatively resistant to policy swings. Hostile federal policy slows deployment but cannot stop it when state policies and market forces point the opposite direction. Supportive federal policy accelerates deployment but isn't the sole determinant of success anymore.

Corporate America: Leading by Example

While the government vacillates, corporate America tends to commit. This creates momentum independent of federal policy and demonstrates how market economics are driving American renewable deployment. Tech companies led the charge in renewables. Post-COVID, Google, Amazon, Microsoft, Apple, and Meta signed massive power purchase agreements with renewable developers, effectively guaranteeing markets and enabling project financing.[6,7,8] These weren't symbolic gestures; we're talking gigawatts of capacity commissioned specifically to supply corporate data centers and operations.

But tech companies weren't alone. Walmart, Target, and other retailers installed solar on massive warehouse roofs and parking lots. Auto manufacturers committed to electric vehicles (partly to meet California's emissions standards, which effectively set national standards due to California's market size). Financial institutions set renewable energy targets for their operations and increasingly for their investment portfolios. This corporate leadership accelerated deployment in two ways. First, direct procurement. Companies building or contracting for renewable generation to power their own operations. Second, indirect influence. Corporate renewable commitments created markets that drove costs down, made projects more financeable, and normalized renewable electricity as the default choice rather than an alternative.

Corporate renewable commitments also have political benefits. When major employers in red states benefit from renewable energy (construction jobs building solar farms, lease payments to landowners and tax revenue to rural counties) it becomes harder for politicians to oppose renewables even if they are ideologically inclined to do so. Both Iowa and Texas, bastions of conservative politics, talk a tough fight against renewables, but tacitly allow development.

The Inflation Reduction Act

The Biden era IRA deserves special attention as potentially the most significant U.S. climate policy ever enacted. The bill provided roughly $370 billion in climate and clean energy incentives over ten years through a combination of tax credits, grants, and loan guarantees. The production tax credit for wind and solar provides payments based on actual electricity generated, incentivizing developers to build projects in the best locations and operate them efficiently. The investment tax credit reduces upfront costs for both utility-scale and rooftop solar, making projects more attractive to investors and homeowners.

Strategically, the IRA included domestic content requirements that incentivize manufacturing in the United States. To qualify for maximum tax credits, projects must use panels, turbines, and batteries manufactured domestically (or in countries with free trade agreements). This creates incentives for building manufacturing capacity in America rather than relying entirely on inexpensive Chinese imports. The results were immediate. Manufacturing announcements for battery factories, solar facilities, and wind component production surged after the IRA passed. States competed to attract these factories with additional incentives. The clean energy manufacturing investment total exceeded $100 billion in the year following IRA passage. More than the previous decade combined.

But the IRA was also vulnerable and was quickly disassembled, piece by piece, after the Trump win in 2024. It had been passed through budget reconciliation with zero Republican votes, meaning it has no bipartisan support to protect it from future rollback. When the Republicans won both congress and the white house, they used their newly found power to cancel IRA programs as quickly as they could. The paradox in this action is that the IRA's economic benefits were designed to flow disproportionately to red states and Republican districts to create political constituencies that might protect the bill even when Republican leadership wants to repeal it; but even these incentives could not stand up to the hold Trump held over Congress, so the IRA was disassembled. The mid-term elections in 2026 may determine if there is a price paid for Republican killing construction and manufacturing jobs in the name of protecting their fossil fuel patrons.

Grid Challenges: The Missing Piece

America's renewable transition faces a severe bottleneck as the electrical grid wasn't designed for high renewable penetration and upgrading it is an extremely difficult political and logistical proposition. The Grain Belt Express transmission line, recently targeted by the Trump administration, would bring renewable energy from Kansas to urban centers in the Midwest. Conceived in 2014, by 2025, it has yet to start construction, despite a pressing need for the power. Decade long planning timeframes are typical of these projects.

The U.S. grid is balkanized into three separate systems (Eastern, Western, and Texas), with limited interconnection between them. Within each system, transmission planning is fragmented across multiple regional entities, state regulators, and utility territories. Building new high-voltage transmission lines requires navigating a maze of regulations, permits, and opposition from landowners and local communities. This creates absurd situations where developers want to build renewable projects, customers want to buy renewable electricity, but the projects cannot be built because transmission capacity doesn't exist to move the power where it is needed. Queue backlogs for grid interconnection stretch years long, with projects waiting for approval to connect to the grid even after clearing all other regulatory hurdles. Today, this is the core problem facing American businesses and residential consumers. Some states are addressing this. Texas built extensive transmission to wind-rich West Texas (though it could use more). California is building transmission from desert solar resources to coastal population centers. But overall U.S. transmission investment lags far behind renewable generation growth, creating increasingly severe bottlenecks.

America's hide-bound project delivery problems can be compared to China, where they build high-voltage transmission at scale and speed that makes American efforts look glacial. Europe coordinates transmission across international borders more effectively than the U.S. coordinates between states. America's federalist system, usually praised for allowing innovation through state-level competition, becomes a liability when continent-scale infrastructure coordination is required. The Chinese government's tight, centralized political control has proven superior in these circumstances.

The Unlikely Coalition

Perhaps the most fascinating aspect of America's renewable transition is the coalition supporting it. A strange alliance of environmental progressives, national security hawks, rural economic development advocates, and fiscal conservatives seeking cheap electricity. Environmentalists support renewables for obvious reasons (climate change mitigation, air pollution reduction and ecosystem protection). National security advocates support renewables because they reduce dependence on imported fossil fuels and decrease petrostate entanglements. Rural development advocates support renewables because they bring investment and steady income to agricultural regions facing economic challenges. Fiscal conservatives support renewables because they are now the cheapest electricity source and reduce costs for governments, businesses, and consumers.

This coalition doesn't agree on much else. Their political alignments differ dramatically on most issues. But on renewable energy, they find common ground based on different motivations leading to the same conclusion: accelerate deployment ASAP. This coalition's "big tent" makes American renewable transitions more politically durable than they might otherwise be. Opponents cannot simply paint renewables as a progressive environmental cause when military leaders, farmers, and business-oriented conservatives also support them. The broader the coalition, the harder it becomes to roll back progress.

What the American Model Teaches

America's messy, fragmented, market-driven renewable transition offers several lessons:

- Market economics can drive transitions even with hostile federal policy and regulatory fragmentation. When renewables become cheapest means of generating electricity, they will get built regardless of political environment.
- State-level leadership matters enormously in federal systems. States can act as laboratories of policy innovation, with successful approaches spreading to other states.
- Corporate procurement creates deployment momentum independent of government policy, providing stability through political cycles.

- Political coalitions broader than environmentalists make renewable transitions more durable. When national security, rural economic development, and cost considerations align with climate goals, political opposition weakens.
- Transmission infrastructure is a critical bottleneck that federalist systems struggle to address. This might be the biggest challenge facing U.S. renewable scaling and there is no hope for a quick solution.
- Policy uncertainty affects pace but not direction. American renewable deployment slowed during hostile administrations and accelerated during supportive ones, but the trend line never reversed because underlying economics continued improving.

The American approach isn't elegant. It is not efficient. It creates enormous differences in outcomes. But it is grinding forward and renewable capacity grows every year regardless of which party controls the federal government. Market forces and state-level leadership compensate for federal policy swings.

By 2025, America is on track for renewable electricity to exceed fossil fuel generation within a decade. Not because of any grand national strategy, but because of thousands of individual decisions by states, corporations, utilities, and citizens responding to economic realities and diverse motivations.

A look at Developing Countries

The next chapter examines how developing countries are approaching renewable transitions and discovering that they can leapfrog the fossil fuel development path entirely by going straight to renewable energy. This is where the global impact becomes most profound, because developing countries will add most of the new electricity generating capacity over the next 20-30 years.

CHAPTER 7

The Global South
Leapfrogs Development

The Global South is a term used to describe countries that are mostly South of the equator, have lower economic status and higher rates of poverty. They could be characterized as being second world countries.

In rural Kenya, a village that never had grid electricity now powers lights, phone charging, and small businesses through a solar mini grid. In India, rooftop solar installations are growing faster than anywhere else on earth, bringing electricity to millions who never had reliable power. In Morocco, massive concentrated solar plants generate enough electricity for millions of homes. Across Latin America, wind and solar farms are being built faster than fossil fuel plants despite the region sitting on substantial oil and gas reserves.

The developing world is teaching the developed world the crucial lesson that you don't need to follow the legacy path using fossil fuels. Countries can leapfrog directly from energy poverty to renewable abundance, the same way they leapfrogged from no phones to mobile phones without building landline infrastructure in between.

This chapter explores how the Global South is not following the industrialized world's carbon-intensive development path, but instead charting a cleaner, faster, less expensive route to energy access. This matters enormously as developing countries will install most of the world's new electricity generation capacity over the next three decades. If they follow the fossil fuel path, climate goals become impossible to accomplish. If these countries go straight to renewables, the energy transition accelerates dramatically and emissions peak sooner globally.

The Leapfrog Effect: Mobile Phones as Precedent

Twenty years ago, telecommunications experts assumed developing countries would build landline telephone infrastructure the way wealthy nations did, with copper wires or fiber-optic cables connecting every building to central exchanges, with decades of investment and a slow build-out. Then mobile phones and wireless data networks happened. Developing countries skipped landlines entirely. Why build expensive fixed infrastructure when wireless towers could cover large areas at lower cost? Today, mobile phone and wireless data network penetration in Africa exceeds the landline penetration that Europe took a century to achieve. And it happened in less than two decades.

Energy is following a similar pattern. Why build centralized, billion-dollar fossil fuel power plants, long-distance transmission lines, and distribution infrastructure to every village when distributed solar plus batteries can electrify communities faster and cheaper? The economics are compelling. A solar mini grid can be installed in a rural village in weeks or months. A grid extension from a distant central power plant takes years, requires extensive transmission infrastructure with engineering and permitting, and often isn't cost-effective for small, dispersed populations. For hundreds of millions of people in rural developing regions, distributed solar is simply the fastest, cheapest path to electricity access. This leapfrog effect extends beyond rural electrification. Even in cities with traditional grids, rooftop solar plus batteries increasingly make economic sense compared to grid-supplied electricity from fossil fuel plants, especially in regions with unreliable grid power or high electricity costs.

India: Renewable Superpower in the Making

India is simultaneously one of the world's largest coal consumers and one of the fastest-growing renewable energy markets.[44] This contradiction reveals the complexity of developing country energy transitions. They are not abandoning fossil fuels immediately, but they are meeting new demand primarily with renewables rather than expanding fossil fuel capacity. India is adding renewable capacity faster than most developed countries' total capacity. The country has set the ambitious targets of 500 GW of non-fossil fuel capacity by 2030.

Why is renewable energy booming in a country still heavily dependent on coal? Several factors:

- Economics: Indian renewable energy is now cheaper than new coal plants, and increasingly less expensive than operating older coal plants. With electricity demand growing roughly 6% annually, utilities choose the cheapest option for new capacity, which is solar and wind.

- Air pollution: Indian cities suffer catastrophic air pollution, with much of it coming from coal power plants. Public health concerns create political pressure to shift away from coal. Renewables offer cleaner air without sacrificing economic growth.

- Energy independence: India has substantial coal deposits but is the second largest importer of oil. This creates economic vulnerability and trade deficits. Renewables can be manufactured domestically (increasingly so as India builds manufacturing capacity) and don't require fuel imports. This improves trade balances and energy security.

- Development needs: Hundreds of millions of Indians still lack reliable electricity. Meeting this need quickly and affordably favors distributed solar rather than slow-building centralized coal plants. Next door, Pakistan imported 17 gigawatts of solar panels from China in 2024, proving that the entire region is well suited to solar.

India's path demonstrates that developing countries can increase renewable deployment while still maintaining substantial fossil fuel generation during the transition. Coal use in India isn't declining yet (though growth is slowing), but new capacity is overwhelmingly renewable. Over time, this shifts the generation mix toward renewables without requiring dramatic coal plant closures that would create reliability concerns.

The Indian model (meet new demand with renewables while letting existing fossil fuel plants gradually age out) might be the pragmatic path for many developing countries. It avoids the economic disruption of forced fossil fuel plant closures, preserves the value of existing generating assets, while still achieving the end goal of renewable dominance over time.

Africa: The Distributed Solar Revolution

Africa is experiencing perhaps the most dramatic energy leapfrog of all. Solar mini grids can provide high-quality uninterrupted renewable electricity to underserved villages and communities across Sub-Saharan Africa and be the lowest-cost solution to close the energy access gap on the continent by 2030. The scale is extraordinary. Kenya's mini-grid project, when completed, will electrify 567 public facilities, including secondary schools, health facilities, and administrative offices, powering water pumps for 380 boreholes and giving access to electricity to approximately 277,000 households, or 1.5 million people.[45]

This isn't aid-dependent charity, but market-driven electrification based on economics. Companies like Renewvia, M-KOPA, and d.light Solar deploy solar systems from small household kits to village mini-grids on a commercial basis, often using mobile money payment systems for financing. Customers pay for electricity the same way they pay for mobile phone minutes with small, regular payments that are affordable while allowing companies to recover costs and earn returns.

The model works because of:

- Lower upfront costs: Distributed solar requires less capital than grid extension. A village mini grid might cost $100,000 to $500,000 versus tens of millions for transmission lines from distant plants.

- Faster deployment: Solar systems can be installed in weeks. Grid extensions take years and often don't happen at all in low-density rural areas where utilities cannot justify the investment.

- Pay-as-you-go financing: Mobile money systems (M-Pesa in Kenya pioneered this) allow customers to pay small amounts regularly for electricity rather than requiring large upfront payments. This makes electricity accessible to people with irregular incomes.

- Reliability: In regions where grid electricity is unreliable (frequent outages, voltage fluctuations), solar-plus-battery storage systems often provide more consistent power than grid connections.

- The result: Africa is electrifying in ways that bypass the 20th century model of centralized generation and extensive transmission grids. This isn't just rural electrification, it is a fundamentally different development path that could leave Africa with a more distributed, resilient, and sustainable energy system than regions that industrialized earlier, all at an affordable cost.

Latin America: Renewable Resource Abundance

Latin America has extraordinary renewable resources. Some of the world's best solar potential is in Chile and Argentina, excellent wind resources range from Patagonia to Mexico, and massive hydroelectric capacity is available in the Amazon basin and in Patagonia. The region is capitalizing on these advantages to build renewable-dominated grids despite having substantial fossil fuel reserves.

Chile has become a renewable energy leader, driven primarily by economics. The Atacama Desert has some of the world's best solar resources (high insolation, clear skies, minimal humidity). Solar electricity in Chile is spectacularly cheap with power purchase agreements regularly coming in under $20 per megawatt-hour, cheaper than almost anywhere on Earth. Chile's mining industry (copper, lithium) requires enormous amounts of electricity, and mines are increasingly signing direct contracts with renewable generators to lock in low costs.

Brazil combines massive hydroelectric capacity (the Itaipu Dam alone generates 90 TWh annually) with growing wind and solar deployment. The combination creates a highly complementary system. Hydro provides dispatchable power and storage, while wind and solar provide low-cost generation. Brazil demonstrates how countries with existing hydro infrastructure can integrate high renewable penetration more easily than those without storage capacity.

Uruguay runs on almost 100% renewable electricity and achieved this not through government mandates, but by creating a level playing field and letting renewables compete on economics. Lacking any natural fossil fuels, Uruguay had long subsidized fuel imports to lower the cost of energy for the population. A decision was made in the early 2010's to remove subsidies and allow every new energy project to compete on a level playing field. Renewables won convincingly and today energy delivered by wind, solar and hydro costs about half of what imported fossil fuels would cost.[46]

Latin America shows that countries with fossil fuel reserves (Brazil's offshore oil, Mexico's natural gas, Argentina's shale deposits) still choose renewables for new electricity generation when economics favor it. Resource abundance doesn't doom countries to fossil fuel dependency if renewables are cheaper.

Morocco: Concentrated Solar Power Pioneer

Morocco represents a different renewable development path using massive, state-led projects to create showcase facilities. The Noor concentrated solar power complex is one of the world's largest solar installations, using mirrors to concentrate sunlight and generate steam that drives turbines. With thermal storage, the facility can generate electricity after sunset, providing dispatchable renewable power.

Morocco's renewable ambitions aren't just about domestic needs (though reducing energy imports matters for a country with limited fossil fuel resources). They are also about positioning Morocco as a renewable energy exporter to Europe. The country is developing underwater transmission cables to export solar electricity across the Mediterranean, effectively turning Morocco into Europe's renewable energy supplier.[47]

The business model of developing countries with excellent renewable resources exporting to developed countries with high demand, could become increasingly common. Australia is pursuing similar strategies, planning to export solar electricity to Singapore via undersea cables and produce green hydrogen for export to Japan. The Middle East (Saudi Arabia, UAE) is pivoting from oil exports to plans for green hydrogen and synthetic fuel production using abundant solar resources. Resource-rich countries recognize that 21st century energy exports will be renewable electricity and hydrogen-based fuels, not fossil fuels. Countries who control excellent renewable resources (abundant sun, strong wind, optimal geography) could position themselves as electricity exporters of the future.

Southeast Asia: The Coal Challenge

Southeast Asia presents one of the most challenging scenarios for developing countries pivoting to renewable transition. The region is building coal plants at a pace that threatens global climate goals, as they face rapid economic growth, rising electricity demand, combined with abundant, low cost coal. Decades long generation contracts are difficult to walk back, even when lower cost alternatives are coming to market.

Vietnam is leading the regional renewable transition, driven by rapid solar deployment that exceeded government expectations. The country installed so much solar so quickly that it had to pause new connections temporarily because grid infrastructure couldn't handle the sudden capacity increase. This created challenges but also demonstrated the enormous appetite for renewable investment when policies are supportive and economics are favorable.

Thailand, Indonesia, and Philippines are all increasing renewable deployment, though coal remains significant in their generation mixes and new generation plants are in the pipeline. The pattern is like India. Not abandoning existing coal plants immediately but also meeting new demand with renewables as costs favor them.

The key question for Southeast Asia is this: will aging coal plants be replaced with new coal or with renewables? Current trends suggest renewables, but the outcome depends partly on whether grid infrastructure can handle high renewable penetration and whether energy storage costs continue to decline to address intermittency.

Why Leapfrogging Works: The Cost Equation

The fundamental motivation for developing countries to leapfrog to renewables is economics. For countries without extensive existing fossil fuel infrastructure, building new renewable generation is cheaper than building new fossil fuel generation in most locations.

Consider a country evaluating these two options for new electricity generation:

- Fossil fuel path: Build central power plant (coal or natural gas), construct transmission lines to reach population centers, build distribution infrastructure to reach customers, source and transport fuel continuously. This approach features high upfront capital costs, ongoing fuel costs, long construction times and centralized infrastructure vulnerable to disruption.
- Renewable path: Deploy distributed solar (rooftop, mini-grids, utility-scale), add battery storage as needed. This approach requires minimal transmission infrastructure for isolated mini-grids or leverage existing grids for utility-scale projects. This approach wins because of lower upfront capital costs (especially for distributed deployment), zero fuel costs, faster deployment, more distributed and resilient infrastructure.

For densely populated urban areas with existing grids, the equation might favor connecting to centralized generation (whether fossil or renewable). But for dispersed rural populations without existing infrastructure, distributed solar often wins decisively on economics. And distributed solar can be deployed immediately, while centralized plants require years of planning and construction.

Climate Implications: Peak Emissions Approaching

If developing countries were to follow the industrialized world's fossil fuel development path, global emissions would continue rising for decades. Developing countries account for most of the global population growth and therefore, most of the increase in electricity demand globally over the next 30 years. If that demand were met primarily with fossil fuels, legacy climate goals would become unachievable, even if first world countries meet their individual emission reduction goals. At the inflection point in 2025, it appears that developing countries are meeting new demand primarily with renewables, meaning global emissions from electricity generation are peaking much sooner than predicted. Combine this with China's emissions peak (Chapter 4) and declining emissions in other developed countries, and you can see that the global emissions trajectory bends toward stabilization and decline much faster than seemed possible a decade ago. This is the most important climate development of the 2020s; developing countries leapfrogging to renewables means global emissions peak not in 2040 or 2050 as once feared, but in the mid-2020s. Every year this accelerates reduces cumulative emissions and makes climate stabilization more achievable.

What the Global South Teaches the World

Developing country renewable transitions offer several crucial lessons:

- Countries don't need to follow the fossil fuel path to development. The assumption that industrialization requires fossil fuels is obsolete. Clean energy can power development.
- Distributed generation models work for populations without existing grid infrastructure. The centralized generation model isn't the only path to electrification.
- Mobile money and innovative financing make renewable electricity accessible to populations previously considered too poor to afford electricity. Business model innovation matters as much as technology.
- Countries with excellent renewable resources can position themselves as 21st century energy exporters, replacing hydrocarbon exports with renewable electricity and green hydrogen.
- Economic growth and emissions reduction are compatible. India, Kenya, Morocco, and others are rapidly growing economically while increasing renewable deployment and (in some cases) beginning to peak emissions.
- Speed matters. Distributed solar can be deployed in months or weeks, versus years for centralized fossil fuel plants. For countries desperate to increase electricity access, speed of deployment favors renewables.

The Global South isn't waiting for developed countries to lead the energy transition. They are leapfrogging to renewables because it is the fastest, cheapest, most practical path to development. And in doing so, they are reshaping global energy markets and making the renewable transition truly global rather than limited to wealthy nations.

Looking Forward

As we transition from Part II (global landscape) to Part III (technology), keep in mind that the technologies we'll examine aren't just serving developed countries. They are enabling the Global South to leapfrog centuries of dirty development in a single generation resulting in the most consequential energy transformation in human history, it is happening right now, and it is driven primarily by economics rather than environmentalism.

Looking Forw...

THE TECHNOLOGY REVOLUTION

CHAPTER 8

Solar's Unstoppable Ascent

Stand on a rooftop in Lagos, Nigeria, and you'll see solar panels. Walk through a suburban neighborhood in Las Vegas, Nevada and you will see the same thing. Drive past farmland in Germany, and solar panels share space with crops in agrivoltaic installations. Scan satellite imagery of the Gobi Desert, and you'll see solar farms stretching for miles. Solar photovoltaic technology has become the world's fastest-growing electricity source, and the trajectory shows no signs of slowing.[48]

This chapter explores why solar is ascending to become the dominant electricity generation technology globally, the innovations pushing efficiency higher and costs lower, and the diverse deployment models that make solar viable in virtually every geographic and economic context.

The numbers tell a remarkable story. Global solar capacity exceeded 1,800 gigawatts by the end of 2024, having doubled in just three years. Solar installations in 2024 alone exceeded 400 gigawatts, which is more than the entire global solar capacity that existed in 2017.[4] This isn't incremental growth. Its exponential acceleration driven by continuously falling costs, improving technology and scaling of manufacturing. In late 2025, Australia has so much solar generation that starting in 2026, utility users will get three hours of "all-you-can-eat" electricity for free.[49] Inexpensive solar is the start of an energy abundance that the world has never seen. And we're nowhere near done. Current projections suggest solar could become the world's leading source of electricity by 2033.[50] No other electricity generation technology has this growth trajectory. Solar is becoming what coal was in the 20th century, the dominant source of electrical power.

The Efficiency Frontier: Breaking Physical Limits

Traditional silicon solar cells face a theoretical efficiency limit called the Shockley-Queisser limit, which is about 33% for single-junction silicon cells under typical conditions. For decades, this seemed like an immovable barrier. Commercially available commercial silicon panels topped out around 22-24% efficiency, and incremental improvements became increasingly difficult. The industry was stuck. Then came perovskite solar cells. A new technology that could be layered on top of silicon to create tandem cells that exceed the single-junction limit. The best performing perovskite tandem cells achieved an impressive 34.85%, surpassing the Shockley-Queisser limit of a single junction silicon solar cell.

Perovskite solar cells are not just a laboratory curiosity. Multiple manufacturers are now producing perovskite-silicon tandem cells approaching or exceeding 30% efficiency (about 50% more efficient than the panels you could buy two years ago). As of June 2024, Chinese manufacturer LONGi holds the world record for perovskite–tandem solar cell efficiency, achieving 34.6% efficiency with a two-terminal device.[51]

Why does efficiency matter when solar panels are already cheap? Several reasons:

- Space constraints: Higher efficiency means more power from limited roof space or land area. For rooftop solar where space is restricted, a 30% efficient panel generates 40-50% more electricity than a 20% efficient panel in the same area.
- Installation costs: Whether installing a 20% or 30% efficient panel, labor costs are similar. Higher efficiency means more power per

panel, fewer panels, which reduces the total system costs, even if the panels themselves are more expensive.

- Transmission and land use: For utility-scale solar farms, higher efficiency means generating more electricity from less land (and leasing less land), reducing environmental impact and making projects viable in land-constrained regions.

The efficiency improvements aren't stopping at perovskite-silicon panels. Researchers are exploring triple-junction cells combining multiple materials to capture even broader spectrums of sunlight. Theoretical efficiency limits for these multi-junction cells exceed 50%. More than twice current commercial technology.

What matters most is that efficiency improvements don't require scientific breakthroughs anymore. They are following predictable engineering optimization paths. Each generation of perovskite cells lasts longer, costs less to manufacture, and converts sunlight more efficiently. This is Wright's Law at work again. More production leads to more learning leads to better, cheaper products, which leads to more deployments.

Building-Integrated Photovoltaics: Solar Everywhere

The next frontier isn't just putting solar panels on roofs; it is making the building itself generate electricity. Building-Integrated Photovoltaics (BIPV) turn windows, walls, and roofing materials into electricity generators. Solar windows use transparent or semi-transparent photovoltaic materials that generate electricity while still allowing light through. Early versions were inefficient (3-5%), but newer technologies are approaching 15% efficiency while maintaining good transparency. Imagine office towers where every window generates electricity. The potential is enormous as modern buildings have far more window area than roof area.

Solar roofing shingles look like regular roof shingles but generate electricity. Tesla's Solar Roof is the most visible example, but many manufacturers now produce solar shingles that aesthetically match traditional roofing while generating power. When replacing a roof anyway, the additional cost for solar is much less than the cost of a new roof plus separate solar panels.

BIPV matters because it turns every new building into a potential power plant. Current solar deployment mostly involves dedicated installations. These are panels on existing roofs or ground-mounted arrays. BIPV makes solar the default for new construction, embedding electricity generation into building design from the start.

The cost equation continues to shift, continues to improve. When BIPV replaces traditional building materials rather than adding to them, the incremental cost becomes minimal. A solar window that costs 20% more than a regular window but generates electricity for 25 years is obviously worth it. Solar roofing shingles that cost 30% more than premium roofing tiles but produce electricity become economically rational. We're approaching a future where it would be economically foolish to build without integrated solar. Why use regular glass when solar glass costs minimally more and generates electricity? Why install regular roofing when solar roofing pays for the premium through electricity generation?

Floating Solar: Utilizing Unused Space

Land constraints limit solar deployment in densely populated regions. Concerns of consuming agricultural land arise when farmland converts to solar farms. Then someone realized, why not put solar panels on water? Floating solar (also called "floatovoltaics") involves mounting solar panels on floating platforms on reservoirs, lakes, canals, wastewater treatment ponds, and even ocean surfaces. The concept sounds unusual but offers multiple advantages:

- No land use conflict: Water surfaces, especially artificial reservoirs and industrial water bodies, aren't competing with agriculture or development. Using these surfaces for solar sidesteps land use debates entirely.

- Cooling effect: Solar panels operate more efficiently when cooler. Floating on water provides natural cooling, increasing electricity generation by 5-10% compared to equivalent land-based installations in hot climates.
- Reduced water evaporation: Solar panels shade the water beneath them, reducing evaporation from reservoirs. In drought-prone regions, this water conservation benefit alone can justify floating solar installations.
- Improve health and yields at fish farms: Fish farm ponds covered with solar cells are cooler and help reduce algae blooms. Global warming is increasing evaporation at ponds, but shade from the panels helps mitigate this.
- Reservoir dual use: Hydroelectric reservoirs can host floating solar arrays, creating hybrid generation sites that combine dispatchable hydro with cheap solar. When the reservoir is full, solar generates while hydro is curtailed. When reservoir levels drop, hydro takes over. A naturally elegant solution.

The scale is growing rapidly. China leads with multiple gigawatt-scale floating solar installations on mining subsidence lakes and reservoirs. India is deploying floating solar on canal networks (also providing shade that reduces evaporation). Singapore, with extreme land constraints, is pursuing floating solar aggressively, including offshore installations. California is trialing several designs on their massive canal network through the central valley.

The technology is still maturing. Challenges continue, including the fact that saltwater environments are harsher than freshwater, requiring more durable materials and raising costs. But costs are falling as deployment scales, and the advantages for land-constrained regions make floating solar economically compelling despite higher installation costs.

Agrivoltaics: Farming and Power Generation Together

Another solution to land use concerns is to combine agriculture with solar generation. Agrivoltaics (or agrivoltaic farming) involves mounting solar panels above cropland at heights that allow farming or grazing beneath them. Done correctly, this creates synergies benefiting both agriculture and electricity generation. The panels provide partial shade, reducing water evaporation from soil and create cooler microclimates. For heat-sensitive crops in hot climates, this shade improves yields rather than reducing them. The electricity generation provides additional income for farmers, making marginal farmland economically viable even in poor growing years. Grazing animals (sheep particularly) can graze beneath and around solar panels, creating dual land use. Pastoral agriculture and electricity generation on the same land. This works especially well with ground-mounted utility-scale solar where there's substantial space between panel rows.

The economics are compelling. Farmland rent for solar projects typically provides more reliable income than crop yields, especially in drought-prone or marginal agricultural regions. With agrivoltiacs farmers can continue some agricultural activities while earning steady payments from electricity generation. This makes renewable energy popular in rural areas that might otherwise be skeptical. It is economic development, not just environmental policy. Money almost always trumps everything else.

Research is ongoing into optimal configurations: what panel heights, spacings, and orientations maximize combined agricultural and electrical output. Early results suggest that for many crops, the optimal configuration provides 70-90% of the agricultural yield while generating substantial electricity. For farmers, 80% of crop yield plus electricity payments easily exceeds 100% of crop yield alone. Agrivoltaics also addresses a political concern: food versus energy competition. If solar farms replace productive farmland, critics can argue we're prioritizing electricity over food. Agrivoltaics sidesteps this by producing both. The political value of this shouldn't be underestimated. It makes solar deployment palatable in agricultural regions where opposition might otherwise be fierce.[52]

The Distributed Solar Revolution

Perhaps the most transformative aspect of solar technology is that it works at any scale. You can have a small rooftop system on a house, a medium-scale, rooftop installation on a commercial building, or a gigawatt-scale solar farm. All use essentially the same core technology (solar panels, inverters, mounting systems). This "Lego block" scalability enables distributed generation in ways impossible with other technologies. Try building a "small" coal plant or nuclear reactor. It doesn't work. Nor would anyone want one of these in their neighborhood. These technologies only make economic sense at large scales, and then you burden the project with the cost and complexity of transmission lines. Natural gas can be scaled down but still requires fuel supply infrastructure, which is very expensive. Solar works economically from a single panel to gigawatt farms. This flexibility is revolutionary.

Distributed solar eliminates transmission losses. Rooftop solar generates electricity right where it is consumed, whereas traditional power plants lose 7% of their electricity in transmission and distribution.[53] In regions with aging, inefficient transmission infrastructure, distributed generation avoids these losses entirely while also avoiding the massive costs of transmission upgrades. Distributed solar also increases grid resilience. A centralized grid vulnerable to single-point failures (one hurricane knocking out a major plant or transmission line) becomes more resilient when generation is distributed. If one neighborhood's solar goes down, it doesn't affect others. During disasters, buildings with solar-plus-battery systems can often maintain power when the central grid fails.

The growth in distributed solar is explosive, with particularly rapid growth in Australia, California, and across developing countries where grid-supplied electricity is expensive or unreliable. Community solar projects (shared solar installations providing power to multiple customers) are enabling renters and people without suitable roofs to access solar benefits.

The Cost Curve: No End in Sight

Solar costs have fallen 90% since 2010.[4] That's a stunning decline that made solar competitive with fossil fuels across the planet, but what many don't realize, is that the cost decline isn't slowing. Solar is still dropping 20% per doubling of cumulative capacity, and we're adding capacity so rapidly that doublings occur every few years. What does this mean practically? Solar that's already cheaper than fossil fuels will become even cheaper. Regions where solar is marginally competitive now will see it become overwhelmingly competitive within years. Applications where solar is currently uneconomic (very high latitudes with limited sun, densely shaded urban areas) will become viable as costs fall further and solar will be introduced in building products.

The implications of solar's ascent are unprecedented in the history of renewable energy. We're approaching a world where the question isn't "where can we afford solar?" but "where wouldn't we install solar?" The default assumption will flip from "use grid electricity" to "why aren't you generating your own power?" This psychological shift is already happening in sunny regions with high electricity costs. It will spread to less optimal locations as costs continue falling.

In the 1950's, Lewis Struass, the Chairman of the American Atomic Energy Commission used the term "too cheap to meter" to describe a utopian future where nuclear reactors provided an endless supply of low-cost electricity.[54] Unfortunately, nuclear became more and more expensive through each iteration and today is one of the most expensive means of creating electricity. Solar power, on the other hand, actually holds the promise of energy that can be very low cost or even free. As pointed out earlier, solar power is so abundant in parts of the Australia that it is free for three hours a day for everyone. Nothing will drive the coming collapse of fossil fuel usage like the plummeting cost of solar power.

Why Solar Wins: The Fundamental Advantages

Solar's ascent to becoming the dominant electricity source stems from fundamental advantages that other technologies cannot match:

- Modularity: Works at any scale from watts to gigawatts. No other generation technology has this flexibility. Single solar panels, built in gigafactories at huge scale, can equip a garden shed or a massive utility array in the desert.
- No fuel required: Zero ongoing fuel costs. Installation costs are the only major expense and after that, sunlight is free.
- Minimal maintenance: No moving parts in solar panels. Maintenance consists mostly of occasional cleaning and inverter replacement every 10-15 years.
- Rapid deployment: Solar farms can be built in months versus years for fossil fuel plants. Rooftop solar can be installed in days.
- Distributed generation: Solar can generate power where it is consumed, eliminating transmission losses and increasing resilience.
- Declining costs: Following Wright's Law learning curve with no end in sight. Every doubling of capacity makes the next installation cheaper.
- Geographic reach: While solar is better in some locations than others, it is viable almost everywhere humans live. Even Germany with modest sun has massive solar deployment because it is still economically rational.
- Low environmental impact: Once installed, solar panels generate electricity with no emissions, no pollution, and minimal ongoing environmental impact.
- Long lifespan: Modern panels are warrantied for 25-30 years and typically continue generating (at slightly reduced efficiency) for 40+ years.

These advantages compound. Modularity enables distributed generation, which increases resilience while reducing transmission needs. Zero fuel costs mean predictable long-term economics. Rapid deployment means responding quickly to electricity demand changes. No other technology combines these many advantages.

The Path Forward: Terawatts to Come

The innovations discussed in this chapter, such as higher efficiency perovskite cells, building-integrated photovoltaics, floating solar, agrivoltaics, are all moving from experimental to commercial deployment. Each breakthrough expands the realm of where and how solar can be deployed, accelerating the transition beyond current projections.

Perhaps most importantly, solar has achieved the critical threshold where further deployment accelerates automatically. Every installation drives costs down through learning curves. Lower costs drive more installations. More installations create more manufacturing scale, more installation expertise, more financing options, more political constituencies supporting continued growth.

The solar revolution isn't coming, it is here, and it is accelerating. No other electricity generation technology has this momentum. In 20 years, our grandchildren will look at solar panels the way we look at electrical outlets: ubiquitous, unremarkable, essential infrastructure that's simply part of how buildings work.

The sun shines everywhere, at least some of the time. And increasingly, wherever the sun shines, humans are capturing that light and turning it into electricity. It is the most abundant energy source we have, and we're finally learning how to harvest it at scale.

A Look at the Rise of Wind Power

Our next chapter examines the rise of wind power to one of the staples of renewable energy. From the storm-tossed North Sea to the plains of Texas and the pampa of Uruguay, wind turbines now provide some of the most reliable sources of electricity and often provide that power 24 hours a day.

CHAPTER 9

Wind: From Niche to Mainstream

D rive across the Great Plains of America, and you'll see them everywhere. Massive white turbines with blades longer than commercial aircraft wings, slowly rotating against the horizon. Fly over the North Sea, and you'll see offshore wind farms stretching across thousands of square miles, each turbine a marvel of engineering operating in one of the harshest environments on Earth. Wind energy has transformed from a niche renewable source into a mainstream electricity generation technology that's competitive with fossil fuels across much of the globe.

This chapter explores how wind power became a major electricity source, the innovations making turbines larger and more efficient, the promise of offshore wind in deep waters, and why wind complements solar in ways that make a renewable-dominated grid achievable.

Global wind capacity exceeded 1,100 gigawatts by the end of 2024, with over 117 gigawatts added in just that year. Offshore wind installed capacity reached 83 GW as 2024 proved to be a record year for construction and auctions.[57] Wind now provides significant percentages of electricity in numerous countries: Denmark gets over 50% from wind, Ireland nearly 40%, Uruguay over 40%. These aren't small, isolated grids, they are industrialized nations proving wind can provide reliable electricity at scale.

The trajectory for wind mirrors solar's explosive growth, though from a larger baseline. Wind technology is mature and proven, but continuing innovations are pushing capacity factors higher, costs lower, and enabling deployment in regions previously considered unsuitable for wind power.

Turbine Gigantism: Bigger Is Better

In the 1980s, a large wind turbine had a rotor diameter of 15-20 meters and generated 50 to 75 kilowatts. Modern land-based turbines have rotor diameters of 150 to 180 meters (nearly two American football fields) and generate 5 to 7 megawatts. Offshore turbines have grown even larger. In June 2024, commercial projects began using 16 MW turbines, and 20+ MW prototypes are rolling out in Europe and China, with Chinese manufacturers recently installing what may be the world's biggest floating wind turbine with rotor blades spanning over 260 meters (almost the height of a 90-story building). Why the relentless drive toward larger turbines? Physics and economics. Wind power scales with the swept area of the rotor (πr^2), so doubling rotor diameter quadruples the swept area and roughly quadruples power output. But manufacturing and installation costs don't quadruple, they increase roughly linearly with size. This means larger turbines generate more power per dollar of investment.

Larger turbines also capture steadier, stronger winds at higher altitudes. At 100 to 150 meters above ground (typical hub height for modern turbines), wind speeds are higher and more consistent than at 50 to 80 meters (older turbine heights). Higher capacity factors (percentage of time generating at rated capacity) mean more electricity from the same installed capacity, improving project economics.

There are practical limits to turbine size. Blade length is constrained by transportation logistics. The blades are too big for current aircraft, too heavy for helicopters and how do you move a 120-meter [nearly 400 feet] blade on roads? The materials they are made of have limits, and very large turbines require specialized installation vessels. But these limits keep expanding as technology advances. The turbines that seemed impossibly large a decade ago are now standard. Twenty years from now, today's "giant" turbines will seem modest.

Offshore Wind: The Massive Untapped Resource

Offshore wind represents one of the largest untapped renewable energy resources globally. Wind over oceans is stronger, steadier, and more consistent than over land. There is enormous space available without land use conflicts. And offshore wind can be deployed relatively close to coastal population centers where electricity demand is highest.

But offshore wind comes with a few challenges. Installation is expensive, hostile marine environments are tough on equipment, and the grid connection requires thick undersea cables. For years, offshore wind was interesting but land-based wind or other generation sources were the low-hanging fruit. But that has changed. Offshore wind costs have fallen dramatically as turbines grew larger while installation techniques improved. By 2024, offshore wind was cost-competitive with fossil fuel generation in many markets, and costs continue declining.

Europe leads in offshore wind deployment, with the North Sea hosting massive wind farms generating electricity for the UK, Germany, Netherlands, Denmark, and Belgium. These installations demonstrate that offshore wind works at large scale. The UK, for example, gets over 12% of its electricity from offshore wind, and that percentage is rising every year as more projects come online.

China is deploying offshore wind even faster than Europe. Chinese manufacturers produce the turbines domestically, bringing costs down through scale. China's long coastline adjacent to major population and industrial centers makes offshore wind particularly attractive. The country is targeting 1,300 GW of offshore wind capacity by 2030 and 2,000 GW by 2035.[58]

The United States started deploying large-scale offshore wind deployment, during the Biden Administration, after years of delays. The East Coast, particularly from Massachusetts to Virginia, has excellent offshore wind resources close to major population centers. The Biden Administration announced a goal of 30+ GW of offshore wind by 2030, but as of the writing of this book, the Trump administration is doing everything possible to derail, delay or cancel these projects.[59]

Offshore wind's advantage over solar is seen in higher capacity factors. Good offshore wind sites achieve 50% to 60% capacity factors (generating at that percentage of rated capacity), compared to 25% to 35% for solar. This means offshore wind generates more electricity from the same potential (nameplate) capacity, making it valuable for grid operators seeking consistent, efficient, 24-hour power generation.

Floating Platforms: Deep Water Wind

Traditional offshore wind turbines are fixed to the seabed with fixed foundations. This works in waters up to about 50 to 60 meters deep. Beyond that, fixed foundations become prohibitively expensive. This limits offshore wind to relatively shallow waters, typically within 30 to 50 km of shore.

Floating wind platforms remove this limitation. Turbines are mounted on floating structures anchored to suction caissons (anchors) in the seabed with cables, allowing deployment in waters hundreds of meters deep. This opens vast offshore areas to wind development, particularly along coastlines where waters deepen quickly (West Coast of North America, Norway, Japan, parts of Mediterranean). While still early-stage, floating wind is commercializing rapidly. Norway's Hywind Scotland, operational since 2017, demonstrated that floating wind works reliably even in harsh North Sea conditions.[60] Larger floating wind farms are now under construction in South Korea, Scotland, and off the coast of California.

Floating wind's economics are improving as designs are refined, and installation techniques mature. Current costs are higher than fixed-bottom offshore wind, but the gap is narrowing and floating wind accesses prime wind regions that fixed-bottom cannot reach, particularly important in countries like Japan, where shallow waters are limited but deep-water wind resources are abundant. The technology is diverse and still developing. Multiple floating platform designs are being tested including spar buoys, semi-submersibles, tension-leg platforms. This design diversity is healthy. It is unclear which approach will dominate, and competition drives innovation. Unlike solar or land-based wind where designs have largely converged, floating wind is still figuring out optimal configuration technology.

Onshore: Wind in Unexpected Places

One surprising aspect of wind's development is that it is economically viable in places that don't seem particularly windy. Iowa gets 60% of its electricity from wind, despite being a Midwestern agricultural state, not a coastal region with obvious wind resources.[41] Texas leads the nation in wind generation, driven by West Texas and Panhandle wind resources that aren't exceptional by global standards but are consistent enough to be economically viable.[61] This is a crucial lesson; wind doesn't require premium locations to be economically viable. Good (not great) wind resources combined with low land costs, supportive policies, available transmission and modern turbines create viable economics. The requirement isn't "best in class" wind, it is "good enough wind plus supporting economics." And available transmission.

"Good enough" expands wind's geographic reach enormously. If only the windiest locations were viable, wind would be limited to specific coastal regions and mountain passes. But if "moderate but consistent" wind resources work economically, then wind becomes viable across vast areas such as the Great Plains, large parts of Europe, extensive regions of Asia, Australia, and Latin America.

Wind and Solar Working Together: The Perfect Pair

Perhaps wind's greatest value is that it complements solar rather than competing with it. Wind often blows strongest at night and during winter, exactly when solar generation is lowest. Solar generates strongest during summer days, when wind tends to be weaker in many regions. This complementarity creates a more stable renewable generation profile than either source alone. Consider a grid with 40% solar and 30% wind versus one with 70% solar. The mixed system has less extreme variability. When sun sets, wind often picks up. When winter reduces solar output, wind generation typically increases. The combined system requires less storage and backup capacity than a solar-only or wind-only system with equivalent total renewable capacity. The only requirement is that both generation assets reside, more or less, in the same transmission area and have available transmission capacity to support the mix of services.

This symbiotic relationship is particularly valuable in regions with strong seasonal variations. Northern Europe has long, dark winters where solar contributes little, but winter storms provide excellent wind generation. Summer brings abundant solar from long days, but less wind. The combination provides relatively steady renewable generation year-round, even before adding storage. The symbiosis isn't perfect everywhere; it depends on local climate patterns. But across much of the world, wind and solar peak at different times, making their combination more valuable than the sum of their individual contributions. Grid operators recognize this, which is why renewable deployments increasingly use strategies that include both technologies rather than focusing exclusively on whichever is locally strongest.

Hybrid Wind-Solar Installations: Maximizing Infrastructure

A growing trend is co-locating wind and solar at the same site, sharing grid connection infrastructure. This combined approach reduces costs (one grid connection, one substation, shared transmission), speeds deployment (less need for new transmission capacity), while maximizing generation from the site. The economics are compelling. Grid connection costs (transformers, switchgear, transmission lines) are often 10-30% of total project costs. Sharing these between wind and solar installations reduces per-megawatt costs significantly. And because wind and solar peak at different times, the same grid connection carries more total energy than if it served only wind or only solar.

Battery storage is increasingly being added to hybrid wind-solar installations, creating triple-hybrid systems. Solar generates during the day, wind at night, and batteries smooth the transitions. These triple-hybrid projects can offer firm capacity, guaranteeing electricity delivery regardless of weather conditions, making them competitive with fossil fuel plants for grid reliability.

Repowering: The Second Wave

Early wind farms from the 1990s and 2000s used small turbines by modern standards, 1 to 2 MW capacity, small rotors, modest hub heights. Many of these turbines are reaching the end of their design life and need to be replaced. But the sites often have excellent wind resources that were only partially utilized by early technology.

Repowering involves replacing old, small turbines with modern, large ones. A site with 50 old 1.5 MW turbines (75 MW total) might be repowered with 10 to 12 modern 7 MW turbines (70 to 85 MW), maintaining similar total capacity while dramatically reducing turbine count. The economics are attractive. The site already has grid connection, access roads, maintenance facilities, and proven wind resources. Permitting is often easier (it is an existing facility, not a greenfield development). And modern turbines generate far more electricity from the same land area.

Repowering is accelerating in Europe and North America as first-generation wind farms age out. This creates a "second wave" of wind growth, not just new sites being developed, but existing sites having capacity upgraded. The potential is significant as repowering early wind farms with modern turbines could add 50 to 100+ GW of new capacity globally without requiring new sites.

The Supply Chain Challenge

Wind's rapid growth creates supply chain strains. Manufacturing turbines, blades, towers, and nacelles (the housing containing the generator and gearbox) requires specialized facilities and skilled labor. Offshore installation requires specialized vessels and there is a global shortage of ships capable of installing the latest generation of massive offshore turbines.

Blade manufacturing is particularly constrained. Modern blades are 80 to 120 meters long for offshore turbines, requiring enormous facilities and precise manufacturing, akin to manufacturing aircraft. Transporting blades this large is logistically complex. They cannot easily travel on roads or railways, limiting factories to locations with port or specialized transportation access. Radia, an aircraft manufacturing start-up, is building the Windrunner, a huge cargo aircraft capable of transporting very long wind turbine blades directly from the factory to the project site, and it can land on dirt runways.

These supply chain limitations are temporary. When demand exceeds supply, prices rise, making new manufacturing capacity economically attractive. Multiple new blade factories, turbine assembly facilities, and specialized installation vessels are under construction or planned. The constraint is real, but it is being addressed through investment responding to market signals.

The supply chain challenge also drives innovation. Manufacturers are developing modular blades that can be transported in segments and assembled on-site, eliminating transportation constraints. New materials (carbon fiber composites, advanced polymers) allow lighter, stronger blades. Automated manufacturing increases production while reducing costs. Within 3 to 5 years, current supply chain bottlenecks will likely ease as new capacity comes online. This will remove a constraint on wind deployment pace, potentially accelerating growth beyond current projections.

Why Wind Matters: The Firm Capacity Question

Wind's greatest contribution to renewable transitions might be addressing the "firm capacity" challenge. Grid operators need generation they can dispatch (access) on demand, which has traditionally been provided by fossil fuel plants that can burn more fuel when demand rises. Solar struggles to provide this (it is strong midday, weak or zero at night and during winter in high latitudes). Wind, specifically offshore wind with high-capacity factors, provides more consistent generation that's closer to firm capacity. While wind isn't perfectly predictable, modern forecasting is remarkably accurate hours ahead, allowing grid operators to position other resources accordingly. And geographic diversity helps. The wind is always blowing somewhere in a large, interconnected grid.

Combining wind with storage creates genuinely firm capacity. Offshore wind farms increasingly include battery storage that captures excess generation during high-wind periods and releases it when wind drops. This wind-plus-storage combination can guarantee capacity in ways that wind or storage alone cannot.

Firm capacity matters enormously for grid reliability as fossil fuel plants retire. Critics argue that intermittent renewables cannot replace dispatchable fossil fuel plants, but wind (particularly offshore wind with high-capacity factors) plus storage creates dispatchable renewable capacity that can replace fossil fuel plants for grid reliability and do it at a lower economic cost. The technology and economics are already there. It is being deployed at scale now, not waiting for future breakthroughs.

Wind's Continued Ascent

Wind capacity will likely exceed 2,000 GW globally by 2030, potentially reaching 3,500-4,000 GW by 2040 if current growth trajectories continue.[62]

The innovations discussed: giant turbines, floating platforms, hybrid installations and repowering all point toward wind becoming an even larger part of global electricity generation. Combined with solar's explosive growth, wind and solar together could possibly provide 60% to 70% of global electricity by 2040, relegating fossil fuels to backup and niche roles. And it should be noted that the wind turbine life cycle is becoming greener. In 2022, Siemens Gamesa installed the world's first Recyclableblade, addressing the issue of wind turbines going to the landfill at the end of their useful lives.

Wind has gone from niche to mainstream in two decades. It will go from mainstream to dominant in the next two. The transition is driven by the same factors accelerating solar; relentlessly declining costs, improving technology, and economics that make wind the rational choice for new generation capacity across vast geographic areas.

Next Up, Energy Storage Provides the Final Piece

One of the greatest innovations of the last five years has been the commercial availability of affordable battery storage for homes, micro-grids and utilities. Storage makes intermittent renewable energy like solar behave more like a dispatchable coal fired power plant. We will also explore energy storage options such as gravity batteries, green hydrogen and heat batteries.

CHAPTER 10

Storage Changes Everything

For decades, the fossil fuel industry rolled out their trump card against clean energy: "The sun doesn't always shine, and the wind doesn't always blow. How will you keep the lights on?" It was a fair question and for most of renewable energy's history, there was no satisfactory answer. Grid-scale energy storage was expensive, limited, and could not economically handle the daily and seasonal variations in renewable generation. Then came the electric vehicle revolution, the manufacturing ramp up of lithium-ion batteries, and a cornucopia of new storage technologies. By 2025, the intermittency problem that plagued renewables for generations had its solution: Battery storage. Not through a single breakthrough, but through multiple overlapping solutions that together transform how we think about electricity storage and grid management.

This chapter explores the storage technologies making renewable-dominated grids possible, from lithium-ion batteries dominating short-duration storage to emerging long-duration solutions for seasonal variations, and the intelligent systems that orchestrate it all.

Lithium-Ion: The Unexpected Solution

Nobody predicted that automobile batteries would solve the grid storage problem. But that's exactly what happened. The massive manufacturing buildout for electric vehicles drove lithium-ion battery costs down 97% since 1991, with prices continuing to fall 19% for every doubling of cumulative production capacity.

By 2024, lithium-ion battery systems reached a price of $165 per meagwatt-hour (MWh) for utility-scale storage, a price point that made 2 to 4-hour storage economically competitive with fossil fuel peaker plants in most markets. And then one-year later in 2025, grid battery prices collapsed to an LCOE of $65 per MWh in some markets, showing that sometimes Wright's Law is almost a cliff.[63] These are the price thresholds that change everything. When storage becomes cheaper than the alternative (natural gas peaker plants that sit idle most of the time and have an LCOE ranging from $110 to $228), grid operators choose storage.[64] The growth is explosive. Global battery storage deployments exceeded 150 GW in 2024, up from less than 5 GW in 2020.[65] This isn't gradual adoption; it is exponential acceleration as economics cross the financial viability threshold across the globe.

Lithium-ion batteries, which are the most common battery chemistry in BESS (Battery Energy Storage System) installations, excel at short-duration storage, capturing excess solar generation during midday and releasing it in the evening, smoothing wind variations over hours and providing grid stability. What has yet to be resolved is long-duration storage; days, weeks, or months of stored energy to handle seasonal variations or extended low-renewable periods. Fossil fuel advocates have always been able to point to a pile of coal or a tank of oil as storage for the future. Renewables needed something similar. But most storage needs are short duration and BESS systems work well in this application. Most grid challenges involve short term variations over hours, not weeks. Lithium-ion solves 80% to 90% of the storage problem, even if it doesn't solve all of it. And for the remaining 10% to 20% requiring long-duration storage, other technologies and techniques are emerging.

Beyond Lithium: The Next Wave of Battery Chemistry

Lithium-ion dominates today, but multiple alternative battery chemistries are being commercialized that offer different advantages, such as lower costs, longer lifespans, improved safety, or using more readily available materials.

Sodium-ion batteries replace lithium with readily available sodium. Sodium is far more abundant (it is literally in seawater), eliminating supply chain and mining concerns around lithium. Sodium-ion batteries have lower energy density than lithium-ion, making them less suitable for EVs where weight and size matter. But for stationary grid storage where weight is irrelevant, lower energy density is a non-issue if the cost is lower. Chinese manufacturers are already producing sodium-ion batteries commercially, with costs potentially 20% to 30% below lithium-ion. In late 2025, Inlyte Energy, a San Francisco manufacturer of Iron-sodium batteries completed laboratory testing of their first field ready products. Sodium-ion is now ready for primetime.

Iron flow batteries store energy in liquid electrolytes containing iron compounds. The key advantage is that they can be charged and discharged indefinitely without degradation, unlike lithium-ion which gradually loses capacity over thousands of cycles. ESS, an Oregon-based manufacturer said iron flow batteries can be a "fast response" storage technology, suitable for both short-term grid stability and long-duration storage.

Another approach, solid-state batteries, replaces liquid electrolytes with solid materials, potentially offering higher energy density, improved safety (no fire risk from liquid electrolytes), and longer lifespans. Their initial use will be in electric cars and after years of testing, multiple companies are approaching commercialization. The prize is batteries that last 30 to 40 years without significant degradation, which matches the lifespan of solar panels and wind turbines.

It's important to understand that battery chemistry innovation is just beginning of the Wright's Law cycle. Lithium-ion is mature and will continue to improve incrementally. But sodium-ion, iron flow batteries, and solid-state technologies are earlier in their development curves. They'll follow the same pattern with each doubling of production driving costs down 15% to 20%(or more), and they are doubling from much smaller bases, meaning faster percentage growth.

Heat Batteries

Industrial processes have used fossil fuels not just for electrical generation, but to also produce heat for steam, melting metal, drying paint or processing food stocks. A rather recent renewable advance is the creation of heat batteries that use cheap and abundant solar to "charge up" a large ballast of material that retains the heat and disperses it 24 hours a day to keep factories operating.

Oakland, CA based Rondo Energy manufactures an industrial scale heat battery that uses purpose built ceramic bricks to build an enormous heat battery. A 100 MWh heat battery can charge up in six to eight hours and then disperse heat over 24 hours to create steam, hot air or charge bake ovens. If you have camped outdoors, you know that the rocks lining the fire pit will often still be warm in the morning, long after the fire has gone out. The ceramic bricks in the Rondo battery operate in exactly the same fashion.

One of the early customers for the Rondo Battery is Holmes Western, ironically, an oil company in California. The heat battery produces steam and replaces a similar system fueled by natural gas. The Rondo solution runs off an on-site solar array and eliminates the variability of the price of natural gas, while providing a cleaner solution. Another customer is Heineken in Portugal, who powers one of their breweries with steam from a Rondo heat battery.

Long-Duration Storage: Solving the Seasonal Problem

Short-duration storage (2 to 8 hours) solves most grid challenges, but truly renewable-dominated grids need solutions for longer durations; days to weeks, or even seasonal variations where summer solar abundance must carry over to winter demand peaks in heating-dominated climates.

Several technologies are emerging for long-duration storage:

- Gravity storage is remarkably simple: lift heavy weights when electricity is cheap and plentiful, lower them to generate electricity when needed. Multiple approaches exist, from lifting concrete blocks with cranes, to pumping water uphill into reservoirs and moving weights up and down abandoned mine shafts. The advantages are minimal degradation, long lifespan, uses simple materials (concrete, water, steel cables). The disadvantage is that this solution requires specific geography (hills for pumped hydro, or deep shafts for weight-based systems). Fortescue operates three iron mines in the Pilbara region of Australia. They are developing a battery-electric train to haul ore to the coast for exporting. Because the run to the coast is downhill, the electric locomotives can charge their batteries using regenerative braking when the train is heavy and then return to the mine on battery power when empty and light. The train never needs charging and consumes no fuel, like a mythical perpetual motion machine. This is a great example of a gravity battery in action.

- Compressed air energy storage (CAES) uses excess electricity to compress air into underground caverns or purpose-built tanks. When electricity is needed, the compressed air is released through turbines to generate power. Modern "adiabatic" CAES captures the heat from compression and uses it when decompressing, significantly improving efficiency. CAES can store energy for days or weeks with minimal losses, making it suitable for seasonal storage.

- Green hydrogen converts excess renewable electricity into hydrogen through electrolysis, stores the hydrogen, and reconverts it to electricity through fuel cells or hydrogen turbines when needed. The efficiency is poor, roughly 30% to 40% round-trip (compared to 85% to 90% for batteries), but hydrogen can be stored indefinitely in enormous quantities, making it viable for seasonal storage despite efficiency losses. A good example of green hydrogen used as storage is the IPP Renewed project in Delta, Utah, that is coming online in 2025. IPP has been the site of a 1.8 GW coal fired generation plant for 40 years. The IPP Renewed project will use new gas turbines to create 840 MW of power that will initially burn natural gas but move to a blend of gas and hydrogen that is created using excess solar capacity. The gas will be stored in salt caverns that exist under the power plant. The goal is to be 100% green hydrogen by 2040. And hydrogen has uses beyond electricity. It can fuel vehicles, power industrial processes, and heat buildings, making it a flexible energy carrier.

- Thermal storage captures heat (or cold) for later use. Molten salt storage, used in concentrated solar power plants, can store heat for 10 to 15 hours. Ice storage (making ice when electricity is cheap, using it for cooling later) provides air conditioning energy storage. District heating systems with large hot water tanks can store heat for days. While these aren't electricity storage, they reduce electricity demand during peak periods by storing energy in thermal form.

The bottom line is that we don't need one storage technology that does everything. We need a portfolio of technologies optimized for different durations and applications. Lithium-ion for seconds to hours. Flow batteries or compressed air for hours to days. Hydrogen or pumped hydro for days to months. Combined, these create a complete storage solution that makes renewable-dominated grids viable.

The Economics of Storage: Crossing the Threshold

The storage game changed when batteries became cheaper than the alternatives they replace. Consider a grid that needs peaker capacity (generation that rarely runs but must be available during highest demand hours.) Traditionally, this meant natural gas peaker plants that sit idle 90-95% of the time. These plants are expensive to build ($1,000-

1,500 per kilowatt), require ongoing maintenance, they need staff round the clock and when they do run, they need fuel. Compare peaker plants to battery storage paired with solar. The solar-plus-storage system costs roughly $1,200 to $1,600 per kilowatt (including both solar panels and 4-hour battery), generates free electricity whenever the sun shines, stores excess for peak demand periods, requires minimal maintenance, no on-site staff and has no fuel costs. The economics favors this type of storage in most markets now, and the gap widens every year as battery and solar panel costs continue falling.

This economic crossover is accelerating storage deployment without subsidies or policy support. Grid operators choose storage today because it is cheaper and more flexible than alternatives, not because regulations mandate it. This market-driven adoption is what makes the storage solution unstoppable. It is indifferent to political elections. The economics improve further when storage provides multiple revenue streams. A battery system can store excess solar, provide grid stability services (frequency regulation, voltage support), push out transmission upgrades by reducing peak loads, and participate in electricity markets by buying low and selling high. These multiple streams of revenue make storage projects highly profitable even as storage costs continue declining.

Vehicle-to-Grid: The Distributed Battery Fleet

The electric vehicle fleet will eventually represent more storage capacity than the entire grid needs for renewable integration. EVs are essentially batteries on wheels. Most are parked 95% of the time. What if they could provide grid services while parked? Vehicle-to-grid (V2G) technology enables EVs to discharge power back to the grid during high-demand periods. A typical EV has 80 to 100 kWh of storage. Park it at home or work, plug it in, and it becomes part of a massive, distributed battery system. The owner gets paid for providing grid services, the grid gets flexible storage, and the EV can still be charged when the owner needs to drive.

The potential scale of V2G is stunning. If 100 million EVs (about a third of the project global EV fleet in 2030) each provide 20 kWh of available capacity to the grid, that's 2,000 GWh of storage, which is more than all dedicated grid storage currently installed globally. And this storage requires no additional manufacturing because it is built into vehicles people are buying anyway for transportation.

V2G is still early-stages. Technical standards are being finalized, business models are being tested, and to be honest, most current EVs don't have bidirectional charging capability. But the technology works, pilots are expanding, and newer EVs are increasingly including V2G hardware. By 2030, V2G could provide a significant fraction of grid storage needs at essentially zero additional cost (beyond the bidirectional charger, which adds roughly $500-1,000 to vehicle cost).

The Smart Grid Revolution: Software as Storage

Perhaps the most important storage innovation isn't hardware, it is software. Modern grid management systems can orchestrate thousands of distributed resources with microsecond precision, effectively creating "virtual power plants"(VPPs), through demand flexibility. Consider air conditioning. It consumes enormous electricity during hot afternoons, exactly when solar generation peaks but storage is expensive. Smart thermostats can shift AC usage to cool buildings aggressively when solar generation is high and electricity is cheap, then reduce cooling slightly during peak prices when storage would be discharging. The building's thermal mass effectively becomes energy storage, with occupants experiencing no discomfort.

This demand flexibility extends across countless applications:

- Electric water heaters can heat water when electricity is cheap and then coast on thermal storage during expensive periods.
- Industrial processes with flexible timing can shift to periods of abundant renewable generation.
- EV charging can concentrate during high-solar periods or overnight wind generation.

- Data centers can shift computing loads to times and locations with excess renewable generation.
- Ice makers and refrigeration can make ice or cool products when electricity is cheap, reducing load during expensive periods.

Aggregated across millions of devices, demand flexibility creates gigawatts of "virtual storage" by not holding electricity in batteries but shifting when it is consumed to match when it is abundant. And unlike physical storage with round-trip efficiency losses, demand flexibility is nearly 100% efficient (you're still consuming the same total energy, just at different times).

The software orchestrating this is increasingly sophisticated. Unlike conventional power plants, Virtual Power Plants (VPPs) can communicate with distributed energy resources and allow grid operators to control the demand from end users. An example is a smart thermostat linked to an air conditioning unit that can adjust home temperatures and manage how much electricity the units consume.

Seasonal Storage: The Final Frontier

The hardest storage challenge is seasonal. Storing summer's abundant solar for winter's heating demand. Or ice for cooling in the summer. This requires storing energy for months, not hours or days.

Hydropower may have a future specifically as energy storage or "water battery". Pumped hydro (using excess solar or wind electricity to pump water uphill, then releasing it through turbines when electricity is needed) offers century-long durability that chemical batteries can't match as well as the ability to store energy for months or even years. It is an elegant role as dams become batteries for solar and wind rather than primary generators.

Green hydrogen emerges as a compelling long-term storage solution. Despite poor round-trip efficiency (30% to 40%), hydrogen has unbeatable advantages for seasonal storage:

- Indefinite storage duration with minimal losses
- Enormous storage capacity in underground salt caverns, depleted natural gas fields, or purpose-built tanks
- Multiple uses beyond electricity (transportation fuel, industrial feedstock, building heating, fuel cell standby power, etc.)
- Leverages existing infrastructure (natural gas pipelines can be converted to carry hydrogen)

The economics of seasonal hydrogen storage are counterintuitive. Yes, losing 60% to 70% of the energy in conversion seems grossly inefficient, but if that energy was free or even negatively priced (excess renewable generation that would otherwise be curtailed), losing 60% of it to storage inefficiency doesn't matter. You're converting excess that would have been wasted into storable energy for later use. This is a great example of how the coming abundance of renewable energy is a complete paradigm shift from fossil fuels. There is always an environmental and marginal cost for fossil fuels, but sometimes renewables are literally free.

The combination of short-duration batteries, medium-duration compressed air or flow batteries, and long-duration hydrogen creates a complete storage solution that handles all time scales from seconds (grid stability) to months (seasonal variations). No single technology does everything, but the portfolio together enables 100% renewable grids.

The Storage Solution is Here Today

The intermittency problem that threatened to limit renewable penetration to 30% to 40% of electricity generation has been solved. Not through a single breakthrough, but through a cascade of innovations:

- Lithium-ion batteries solving short-duration storage at costs competitive with fossil fuel alternatives
- Emerging battery chemistries (sodium-ion, flow batteries, solid-state) providing diversity and continuing cost declines

- Long-duration storage technologies (compressed air, pumped hydro, gravity and hydrogen) handling seasonal variations
- Vehicle-to-grid leveraging the massive, distributed storage capacity of EV fleets
- Smart grids and VPPs creating "virtual storage" through demand flexibility
- Software systems orchestrating millions of distributed resources with precision

Together, these solutions remove the technical barriers to renewable-dominated grids. The question is no longer "can intermittent renewables provide reliable electricity?" It is "how fast can we deploy the storage and smart systems to enable 100% renewable grids?"

The fossil fuel industry's trump card has been played and is now played out. It no longer works. As several countries around the world have shown, the lights stay on, the grid stays stable, and industrial processes continue running smoothly, all with renewable-dominated electricity plus storage. This has been demonstrated at scale across multiple grids globally.

Looking Forward

The next chapter looks at peripheral players in the renewables world and how they form an important "supporting cast". Hydro, nuclear, geothermal and fuel cells / hydrogen.

CHAPTER 11

The Supporting Cast: Hydro, Nuclear, Geothermal, and Fuel Cells

Solar and wind will dominate the renewable transition. This isn't speculation, it is what the economics dictate. But understanding 2050's electricity grid requires examining four other technologies that will play supporting roles: hydropower, nuclear, geothermal, and fuel cells. Think of it this way, solar and wind handle 90% of the work while the technologies covered in this chapter manage the remaining 10% by filling specific niches or continuing to operate because they're already built. None of these will drive the renewable revolution, but each provides an important role.

Let's examine the supporting cast through the lens that has guided this book: economics, scalability, and deployment timeline.

Hydropower: The Incumbent That's Staying

Hydropower provides about 16 percent of global electricity today, making it the largest non-fossil contributor. It is dispatchable, long-lasting (50 to 100 years), and has excellent economics for existing facilities at a LCOE of $61/MWh per a French study in 2025.[66] But new large dams cost $150/MWh to build. This is not close to being competitive with solar and wind and finding new build sites is problematic as all the best sites are already developed. Dams displace communities, destroy ecosystems and are increasingly despised.

Climate change is also making existing facilities less reliable as droughts rob the dam of power producing water. With development timelines running 10 to 20 years, dams are also too slow to make a difference for AI and other industries screaming for more power now.

The most realistic scenario is that existing hydropower continues operating and pumped hydro storage (discussed in the previous chapter) expands modestly in favorable locations. Hydropower's percentage of total generation declines from 16 percent today to perhaps 12% to 15% by 2050. Not because hydro fails, but because everything else grows faster.

Nuclear: The Incumbent Facing Decline

Nuclear currently generates 10% of world electricity at extremely high-capacity factors (90 percent or higher) with zero emissions and nuclear has a number of very positive attributes as a source of renewable energy. Operating costs for existing plants run a paltry $35/MWh, so extending their lives to 60 or 80 years or more is economically sensible. They're cheap, they're operating and they offer essentially clean energy.

New nuclear is the problem. Lazard's 2024 analysis shows new nuclear power's LCOE at $140 to $220 per megawatt-hour. Three to five times more expensive than solar ($24 to $96) or wind ($24 to $75). Georgia's Vogtle Units 3 and 4 (online since 2023) cost $35 billion for 2.4 gigawatts, which is roughly $14,500 per kilowatt. Utility-scale solar costs $800 to $1,200 per kilowatt. Large nuclear power plants can cost ten times more to build than renewable alternatives and these projects almost always run over budget and over on schedule.[21]

Unlike solar and wind, nuclear costs have increased over time, not decreased. This is the opposite of Wright's Law. Custom-built one-off projects don't benefit from manufacturing learning curves. Add to that a 10-to-20-year construction timeline competing against renewables getting cheaper every year (and can be deployed rapidly), and the economics collapse. Private capital has abandoned new nuclear in competitive markets, meaning governments must subsidize what investors won't touch.

Small Modular Reactors (SMRs) promise standardized factory production and faster deployment. The problem is that they don't exist at commercial scale yet. NuScale's Utah project was cancelled when costs reached $89 per megawatt-hour and looked to climb above that. Even if SMRs eventually hit their $50 to $60 per megawatt-hour targets, they won't deploy until 2030 to 2035. By then, solar will be even cheaper, and batteries more capable. Clean, inexpensive, fast to deploy nuclear power is something to be desired, but the concept lacks proof today.

The most likely scenario for nuclear is that existing plants operate through the 2060s–2070s. New construction remains minimal except in China. Nuclear energy's global share falls from 10 percent today to 6% to 8% by 2050. Not because nuclear is bad, but because solar, wind, and storage solve the problem at one-third the cost and one-tenth the deployment time, and have no toxic legacy. The outlier would be a successful SMR design that dramatically drops the price and the time to deploy, but keep in mind even a cost-effective SMR design requires staff, maintenance and fuel, driving up the operating costs.

Geothermal: The Wild Card

If any technology here could surprise on the upside, it is geothermal. Traditional geothermal (hot steam near the surface) is geographically limited but excellent where available: 24/7 baseload power with 85% to 95% capacity factors at $60 to $100/MWh.[67]

A drilling technique called Enhanced Geothermal Systems (EGS) has changed everything regarding geothermal. EGS applies gas fracking techniques (multiple, horizontal bores) to hot rock formations anywhere on the planet, creating fractures to increase surface area, and then circulate water to capture the heat. With EGS, geothermal isn't limited to volcanic regions as the earth is hot a few kilometers down almost everywhere. Project Cape Station in

Beaver County, Utah has proven the concept. Fervo Energy drilled hot rock, achieved 3.5 megawatts of flow in 2023, and based on this successful trial, signed power purchase agreements with Google and Southern California Edison. A 400 MW full scale power plant is under construction. The projected LCOE for Cape Station is not available, but projected costs are $45 to $75/MWh for new technology geothermal in the US.[68] Between proof-of-concept in Nevada and production wells in Utah, Fervo tripled power output delivery per well while reducing drilling time by 70 percent.[69,70] Wright's Law is working at Cape Station.

The western United States, East Africa, Indonesia, and even cold regions like Germany and the Arctic could develop significant EGS capacity. Most importantly, geothermal is dispatchable. You can count on it whenever needed. This addresses the "last 10 percent" problem when solar isn't generating, wind is calm, and batteries are depleted. In these cases, geothermal provides firm renewable capacity without fossil fuel backup.

Realistic role by 2050: capacity grows to 100 to 150 gigawatts (roughly 8 to 10 times current levels) and provides 3% to 5% of global electricity. In favorable regions, that percentage reaches 10% to 20% percent. Geothermal won't replace solar and wind, but it could be the perfect complement by providing firm dispatchable baseload where resources exist.

This is the only technology here that might genuinely surprise upward. The learning curve is just beginning, and Wright's Law might make it a winner.

Fuel Cells and Hydrogen: The Efficiency Problem

Hydrogen comes in different varieties based on how it is processed. "Gray" (derived from natural gas and with harmful emissions), "blue" (gray gas with emission controls) and "green" (made from electrolysis using renewable electricity). Understanding the origin of the hydrogen being used will help you understand its ultimate merits as an energy source, but keep in mind only green hydrogen (made from water with renewable electricity) is considered renewable.

Hydrogen can be used in fuel cells to make electricity with only water vapor as emissions. Hydrogen can also be burned as a fuel in vehicles, ships or in turbines to generate electricity via steam. In these cases, while most of the emissions are water vapor, nitrogen oxide is also created. So, a clean fuel in some use cases, and a slightly dirty fuel in others.

Green hydrogen has a few deficiencies relative to other renewables. The first is efficiency. Solar electricity is used to create most green hydrogen production (electrolysis) at 60% to 80% efficiency. Compression for storage or transportation costs another 10%. Fuel cell conversion to electricity is 50% to 60% percent efficient. Total round-trip efficiency is roughly 18% to 46% percent. Batteries, on the other hand, achieve 60% to 80% round-trip efficiency. You lose more than half the energy with hydrogen but only 20% to 40% with batteries. If your application is grid storage, why use hydrogen when batteries are much more efficient?[71] The answer, of course, is duration of storage. Batteries last minutes or hours, hydrogen can be stored for months or even years before use.

Fuel cells as a means of electrical production make sense only in specific circumstances. Microgrids and campus installations could use them for backup power with combined heat-and-power systems capturing waste heat for space heating. Heavy transportation might eventually use fuel cells where refueling speed matters more than efficiency or shipping and where energy for days or weeks is required. But even these niches face battery competition. Lower cost sodium-based batteries and new tech solid state batteries promise longer life and faster charging, potentially eliminating hydrogen's advantages even for heavy transport.

Infrastructure is another barrier. Hydrogen requires entirely new distribution networks or modifications to natural gas pipelines. Building fueling stations costs $1 to $2 million each, on the other hand, electricity distribution already exists everywhere. An alternative for hydrogen is delivery by truck, but again, this introduces cost and lowers the efficiency.

The best guess for 2050 has fuel cells occupying minimal grid roles and delivering less than 1 percent of electricity. Not because the technology is bad, but because efficiency penalties make batteries superior for almost all applications, and infrastructure requirements make deployment expensive and slow.

The Numbers Don't Lie

Comparative economics determine winners and losers. Lazard's 2024 analysis:

Solar PV: $24 to $96/MWh

Onshore wind: $24 to $75/MWh

Geothermal (traditional): $60 to $100/MWh

Geothermal (EGS): $50 to $80/MWh

Nuclear (new): $140 to $220/MWh

Hydrogen fuel cells: $100 to $250/MWh

Solar and wind are dramatically cheaper. Only existing hydropower approaches their economics, and it will be because of these cost factors that solar and wind will prevail not only against fossil fuels, but other renewables for the majority of use cases.[21]

Cost trajectories are almost more important than current price. Since 2010:

Solar costs fell 90 percent

Wind costs dropped 70 percent

Battery costs down 97 percent

Nuclear costs increased roughly 25 percent

Hydrogen electrolysis fell 60 percent but started so high it remains
   expensive

Technologies descending the learning curves (solar, wind, batteries, geothermal) will keep improving. Mature technologies (nuclear, hydro) won't. The gap widens every year. Deployment timelines matter also. Solar farms deploy in 12 to 24 months. Big nuclear takes 10 to 20 years and SMRs, while promising, are still years away as a commercial technology. When solar projects become operational in 18 months, they're built with current technology at current prices. When nuclear takes 15 years, it is locked into yesterday's economics, competing against renewables that improved dramatically during construction.

The 2050 Grid

Based on these economic realities, a typical renewable-heavy grid might look like this in 2050:
  60% to 70% solar and wind
  15% to 20% battery storage
  5% to 10% existing hydro
  3% to 5% geothermal
  3% to 5% existing nuclear
  Less than 5 % everything else

Solar and wind dominate because they're cheapest and fastest to deploy. Batteries excel at short-duration storage with superior efficiency. Existing hydro and nuclear continue because they're already built. Geothermal fills niche roles where it can be drilled. Everything else occupies fractional shares addressing edge cases. This isn't a goal or a desire; it is what the economics dictate. Utilities don't build expensive plants when cheap alternatives exist, and investors don't fund 15-year long construction projects when 18-month alternatives deliver better returns.

Looking Forward

The next chapter examines how smart grids and AI solve the integration challenges that make this renewable future practical and reliable. The renewable revolution isn't just about hardware; it is about digital systems managing complexity at scales impossible for human operators. That management agility is where the transformation becomes truly revolutionary.

CHAPTER 12

The Smart Grid Revolution

Historically, the electrical grid was "dumb". A one-way system where centralized power plants pushed electricity through transmission lines to passive consumers. Grid operators balanced supply and demand by adjusting electrical generation at a few hundred large plants. Consumers were offered fixed price services and couldn't adjust consumption in response to grid conditions. Unfortunately, our electrical grid today is not markedly different from what existed in 1930, but a major upgrade is underway. The modern smart grid is becoming bidirectional, dynamic, and intelligent. Millions of distributed resources generating and consuming electricity, sophisticated software balancing supply and demand in real-time, consumers actively participating in grid management, and AI systems optimizing everything from renewable generation forecasting to demand response. This transformation from dumb to smart grid will be required to make renewable-dominated electricity systems practically achievable and affordably priced.

This chapter explores how grid management is evolving to handle high levels of renewable power generation, the role of AI and machine learning in orchestrating services, the emergence of microgrids that provide resilience, and how electricity markets are adapting to renewable-dominated systems.

From Centralized to Distributed: The Architecture Revolution

The 20th century electrical grid was rigidly hierarchical and centralized by necessity. Large fossil fuel and nuclear plants generated electricity. High-voltage transmission lines moved it long distances. Distribution networks stepped voltage down and delivered electricity to end users. Information flowed only from grid operators to power plants. Consumers were passive recipients. This architecture made sense for the technology available and fossil fuel plants needed to be large to be efficient. Nuclear plants needed enormous scale to justify costs. Renewable energy was expensive and rare. Affordable battery storage didn't exist, and consumers had no way to respond to real-time grid conditions.

Everything about the grid is now changing. Solar panels on millions of rooftops generate electricity "behind the meter" where it is consumed by the owner of the panels but also sent into the grid. Batteries in homes, businesses, and vehicles provide distributed storage. Smart devices can adjust consumption (home air conditioners, for example) in response to price signals or grid conditions. The grid is becoming a distributed network where resources generate, store, and consume electricity dynamically based on always changing, real-time conditions. This architectural transformation requires fundamentally different grid management. Rather than controlling a few hundred large plants, grid operators must install communication systems able to coordinate millions of distributed resources. Rather than one-way power flow, they manage bidirectional flows as "uploaded" distributed generation sometimes exceeds local demand. Rather than static operations, they respond to rapidly varying renewable generation and dynamic demand patterns. The grid is moving from static to dynamic.

AI-Powered Forecasting: Predicting the Unpredictable

The sun doesn't always shine, and the wind doesn't always blow, but modern forecasting systems can predict when they will with remarkable accuracy. AI and machine learning algorithms analyze weather patterns, satellite imagery, historical generation data, and real-time sensor readings to forecast renewable generation hours to days ahead with 95%+ accuracy. This forecasting accuracy transforms intermittency from chaos and surprise into a manageable challenge. If grid operators know 24 hours ahead that wind generation will drop 30% tomorrow afternoon while solar generation will be strong, they can position resources accordingly. They charge batteries during high solar, schedule flexible demand, prepare backup generation if needed and coordinate with neighboring grids for imports or exports.

AI systems now being deployed improve continuously through machine learning and every forecast is compared to actual outcomes, errors are analyzed, and models are refined. What seemed impossibly complex a decade ago (forecasting generation from thousands of distributed solar installations and wind farms) has become routine because AI systems can process enormous data streams and identify patterns humans couldn't detect. Europeans are at the forefront of this technology, as discussed in Chapter 5.

Real-Time Grid Balancing: Microsecond Precision

Traditional grids balanced supply and demand at timeframes of minutes to hours. Operators would adjust generation schedules hours ahead, then make minor adjustments as needed. Modern renewable-dominated grids require balancing at millisecond to microsecond timeframes. Why? Solar generation can drop dramatically when a cloud passes over a large solar farm (seconds to minutes). Wind generation can ramp up or down with changing weather fronts (minutes to hours). Millions of distributed resources change consumption patterns continuously. Managing this requires software systems that can detect imbalances and respond faster than humans could possibly react.

Modern grid management systems use automated controls that:
- Detect frequency deviations (indicating supply-demand imbalance) within milliseconds

- Control battery storage to inject or absorb power within tens of milliseconds
- Adjust flexible loads (water heaters, air conditioners, EV chargers, industrial processes) within seconds
- Coordinate distributed resources across regions to balance local imbalances
- Communicate with neighboring grids to import or export power as needed

This happens continuously, thousands of times per hour, with humans overseeing but not directly controlling most operations. The software responds to conditions faster and more precisely than human operators could, maintaining grid stability that would be impossible to achieve manually.

Microgrids: Resilience Through Independence

While much of the grid is becoming more interconnected, microgrids represent the opposite trend, towards grid independence. These are small, self-sufficient electricity systems that can disconnect from the main grid and operate on a standalone basis. Microgrids in rural Africa were discussed in Chapter 7.

Microgrids in first world countries typically serve a campus, military base, community, data center or industrial facility. They include local generation (usually solar plus storage), local loads, and controls that can manage the internal grid independently. Under normal conditions, microgrids connect to the main grid, exporting excess generation or importing when needed. During grid outages or extreme events, microgrids disconnect and operate independently, maintaining power to critical loads.

The microgrid value proposition is resilience achieved through independence. When hurricanes, wildfires, or other disasters knock out the main grid, microgrids keep essential services running at hospitals, emergency shelters, water treatment plants and telecommunications equipment sites. As extreme weather becomes more common (partly due to climate change), resilience becomes more valuable and microgrids a necessity.

Microgrids are proliferating across the United States, particularly in disaster-prone regions. Puerto Rico, devastated by Hurricane Maria's grid destruction, is installing numerous microgrids to avoid future total blackouts. California is deploying microgrids in wildfire zones where main grid reliability is questionable. Texas, after the 2021 winter storm grid collapse, is seeing surging interest in microgrids for critical facilities.[72]

As battery storage costs fall and distributed solar becomes ubiquitous, the cost premium for microgrid capability shrinks. A hospital that's installing solar-plus-storage for economic reasons can add microgrid controls for relatively modest additional cost, gaining resilience as a bonus. We're approaching a world where microgrid capability becomes standard for critical facilities rather than an expensive upgrade. Even homeowners with solar and batteries can now operate independently from the grid.

Interconnections: Bigger is Better

One of the most effective ways to manage renewable variability is to expand the geographic area over which the grid balances. The wind is always blowing somewhere across a continent. The sun is shining somewhere. Demand peaks at different times in different regions. Large, interconnected grids can leverage these geographic diversities to smooth out variations. This is why Europe's continent-scale grid integration is so valuable. When Germany has excess solar, it exports to France. When wind is strong in Denmark, it exports to Germany. When Norway's hydroelectric reservoirs are full, other countries reduce their hydro generation and import from Norway. This constant balancing across regions makes high renewable penetration easier for everyone.

The United States has three separate grids (Eastern, Western, and Texas) with limited interconnection between them. This balkanized, historical artifact is increasingly problematic. Building more interconnection capacity between these grids would provide enormous value. Texas's wind could serve Eastern demand, Western solar could serve Texas in the evening and Eastern demand could access Western storage. The separate grids are less efficient and require more backup capacity than a unified national grid would need. Building new interconnections is politically and logistically challenging as transmission lines cross multiple states, face local opposition, and require coordinated planning across regions. But the economic value is so large that interconnection expansion is happening despite challenges. Multiple projects are under development to link the Eastern and Western grids, and to better connect Texas to both. Permitting reform is such an issue that even the highly divided American Congress is addressing it at the end of 2025.

The Smart Grid Reality

The smart grid revolution isn't coming, it is here and now just needs to be rolled out widely. Multiple regions globally operate grids with sophisticated AI-powered forecasting, real-time automated balancing, active demand response, and coordination of millions of distributed resources. These systems work reliably, maintaining or exceeding the reliability of conventional grids while integrating high percentages of renewable generation. The transformation from dumb to smart grid happened faster than most predicted, driven by rapidly improving software and declining costs for sensors, communications, and computing. What seemed impossibly complex a decade ago is now standard operations in leading regions. This matters enormously for the renewable transition. The smart grid revolution removes the technical barriers that once limited renewable penetration. Intermittency, variability, frequency stability, voltage support, all the technical challenges that critics claimed would prevent renewable-dominated grids, are being been solved through intelligent software orchestrating distributed resources.

The remaining barriers to 100% renewable grids are no longer technical. They are economic (though economics are rapidly favoring renewables), political (though market forces are overriding political resistance), and logistical (building transmission, deploying storage, upgrading distribution systems). But there's no longer any technical reason preventing grids from operating reliably with extremely high renewable penetration.

Looking Forward

Part III (technology) was about the technical parts of the renewable revolution and now Patt IV (overcoming political and cultural obstacles) discusses forces trying to slow down or kill renewables, including political forces, legacy fossil fuel companies, the myths and issues surround grid reliability and the legitimate concerns surrounding the raw materials needed to make the green grid work.

PART IV

OVERCOMING THE HEADWINDS

CHAPTER 13

Political Opposition and Why It Will Ultimately Fail

In 2023, the Republican-controlled and fossil fuel friendly Texas legislature passed multiple bills designed to slow renewable energy deployment. They legislated charging wind and solar projects special fees, mandating payments to fossil fuel plants for "reliability," and restricting transmission expansion to wind-rich regions. Industry groups warned these bills would devastate Texas's renewable sector, the largest in the nation. Yet in 2024, Texas set records for solar generation, wind production, and energy storage deployment. In that year, 92% of new electrical generating capacity in that state came from renewable projects.[11] Political opposition slowed deployment marginally, but the economics of renewables were so compelling that renewable development continued unabated.

This pattern repeats globally. Fossil fuel interests with deep pockets deploy enormous political influence to slow renewable transitions that threatened their profits. They write and sponsor fossil fuel friendly legislation, fund opposition campaigns, spread misinformation, and lobby aggressively against renewable-friendly policies. And yet, renewable deployments continue to accelerate because the underlying economics overwhelm political resistance.

This chapter examines why fossil fuel industries fight so desperately despite losing battles, the tactics they employ, how political resistance varies across different governmental systems, and why political opposition ultimately cannot stop transitions driven by market fundamentals.

The Fossil Fuel Industry's Last Stand

Understanding why fossil fuel companies fight so fiercely requires understanding what they are facing: The complete obliteration of their business models. This isn't competition from a new product; it is an existential threat to assets worth trillions of dollars today, but potentially worthless in the future.

Consider the situation from a fossil fuel company's perspective. You have:

- Oil and gas reserves in the ground, valued on balance sheets at prices assuming future extraction and sale. If renewables plus EVs eliminate demand for your product, these reserves become worthless, "stranded assets", that can never be profitably extracted and their value evaporates.
- Extraction infrastructure. Drilling rigs, refineries, pipelines, LNG terminals, etc. representing hundreds of billions in invested capital. If this infrastructure becomes obsolete before reaching the end of its expected lifetime, shareholders lose enormous value.
- Expertise and workforce specialized in fossil fuel extraction and processing. Most of these skills don't translate to renewable energy, meaning your human capital depreciates as the industry shrinks. It is interesting, however, that geothermal development may allow these workers to easily pivot to a new job in renewable energy.
- Deep and established political relationships built over decades through lobbying and campaign contributions. This influence evaporates as the industry's production shrinks and can no longer fund political operations at scale. At some point no one returns your calls.

The fossil fuel industry isn't fighting to maintain market share; they are now fighting for their survival. Either they successfully slow the renewable transition long enough to extract remaining reserves and recover infrastructure investments, or they face accelerating collapse as stranded assets multiply and markets disappear. This explains the desperation. Fossil fuel companies aren't irrational, they are cornered. They are fighting because the alternative is accepting trillion-dollar losses and obsolescence. They know they are losing the war, but winning daily battles mean putting off the inevitable by even a few years, and extracting billions more in value before they must sell the corporate jets and close the Michelin rated executive dining room.

The Playbook: Delay, Deny, Confuse

Fossil fuel political opposition follows a predictable playbook, refined over decades of fighting environmental regulations:

- Deny the problem exists. Question climate science and close down publicly funded research. Fund contrarian researchers and amplify doubt about renewable viability. If there's no crisis, there's no need for renewable transitions. This tactic worked for decades but is increasingly ineffective as climate impacts become undeniable and renewables become obviously less expensive at scale.

- Use FUD (fear, uncertainty and doubt). Your lights will go out, you'll die in a blizzard, businesses will close or leave the country, etc. Texas Governor Abbot, blaming wind turbines for deaths during the blizzard of 2021 said, "this shows how the Green New Deal would be a deadly deal for the United States of America". We now know that the deaths in Texas were caused by a network wide failure, including fossil fuel generation sites and nuclear power plants.

- Question the solution. Acknowledge climate change might be real but argue renewables cannot work at scale. They are too expensive, too unreliable, require too much storage, cannot power industrial economies, etc. This tactic is failing as multiple countries demonstrate high renewable penetration with maintained reliability and industrial competitiveness. Even in cold climates.

- Emphasize costs, ignore benefits. Focus relentlessly on renewable deployment costs, subsidies, and rate impacts while ignoring health costs of air pollution, climate damages, or the fact that renewables are now cheaper than fossil fuels. Frame renewable transitions as expensive burdens rather than economically beneficial. "Your power bill is more expensive because of green energy".

- Promote fossil fuels as the "bridge" or "necessary backup." Acknowledge that renewables are growing but argue fossil fuels remain essential for reliability, economic stability, or national security. Position natural gas as "clean" compared to coal (ignoring methane leakage and emissions). This delays fossil fuel phase-out by making it seem more gradual and necessary.

- Protect specific interests. When broader arguments fail, focus on protecting particular constituencies such as coal mining communities facing job losses, petrochemical regions dependent on fossil fuel industries, rural areas hosting fossil fuel extraction. Frame renewable transitions as hurting workers and communities rather than helping them.

- Capture regulatory processes. Influence regulators to slow renewable deployment through permitting delays, interconnection obstacles, and favorable rules for incumbent fossil fuel generators. Regulatory capture requires less public visibility than legislative battles and can be very effective at slowing deployment. This is why oil companies universally backed Trump in the 2024 election. If you are behind in the game, change the referees.

- Use astroturf campaigns. Create fake grassroots organizations with innocuous names to oppose specific renewable projects or policies. These front groups hide fossil fuel funding while creating the appearance of authentic local opposition. A study of "public interest groups" opposing offshore wind in the Northeast revealed that they were almost universally backed by fossil fuel interests.

The playbook is well-documented and increasingly recognized. This makes it less effective as media and green energy activists now routinely identify fossil fuel funding behind apparently grassroots opposition, reducing the tactics' impact. But the playbook persists because it is the only strategy available to industries facing obsolescence.

Why Renewable Opposition Succeeds (Temporarily)

Despite losing the war, fossil fuel interests win many battles. Why? Financial resources remain enormous. Even as the industry shrinks, major oil and gas companies generate hundreds of billions annually in revenue, even while they layoff thousands to "juice" the profits. A small fraction of this money funds substantial lobbying operations, campaign contributions, and opposition campaigns. Renewable industries, while growing, don't yet have comparable financial resources for political operations.

Concentrated benefits, diffuse costs. Fossil fuel industries provide concentrated economic benefits to specific regions (Texas oil, West Virginia coal, Louisiana natural gas). Politicians from these regions face intense pressure to protect fossil fuel interests. Renewable benefits are more diffuse (cheaper electricity for everyone, cleaner air everywhere) which creates less concentrated political pressure.

Existing infrastructure advantages. Fossil fuel infrastructure already exists in the form of power plants, pipelines, refineries. Proposals to shut down existing facilities face opposition from workers, communities, and owners who lose value. The current grid favors big, centralized generation. Building new renewable infrastructure faces opposition from people who don't want it in their area. The asymmetry favors incumbents. Fossil fuel interests connect "new" with "scary".

Regulatory inertia. Regulatory systems designed for conventional generation don't easily adapt to distributed renewable generation. Changing rules requires sustained effort, and fossil fuel interests delay changes through legal challenges, lobbying, and influence over regulators. This inertia slows renewable deployment even when economics favor it.

Short-term thinking. Elected officials face short election cycles and prioritize immediate concerns over long-term transitions. Fossil fuel interests exploit this by emphasizing short-term costs and disruptions while renewable benefits accrue over decades. Politicians avoid imposing near-term costs even when long-term benefits are clear.

These advantages are real but declining. As fossil fuel industries shrink, their financial resources will diminish. As renewable industries grow, their political influence increases. As renewable cost advantages become overwhelming, politicians find it harder to justify protecting fossil fuels. The tide is turning, but it is a gradual process rather than sudden shift.

Regional Variations: How Politics Affects Speed

Political systems and cultures dramatically affect renewable deployment speed, but, crucially, they affect speed rather than direction. Hostile political environments slow transitions, supportive environments accelerate them, but underlying economics drive transitions everywhere eventually. Authoritarian systems with renewable commitment (China) can deploy fastest. Centralized decision-making, state-directed financing, and ability to override local opposition enable rapid deployment. China's renewable buildout proceeds at speeds impossible in democratic systems. But this advantage requires the government supporting renewables. Authoritarian systems opposing renewables (Russia – post Ukraine invasion) can also block deployment just as effectively.

Parliamentary democracies with coalition governments (Germany, Denmark, Spain) deploy renewables rapidly when environmental parties hold influence in coalitions. Multi-party systems allow green parties to punch above their electoral weight by being necessary coalition partners. But coalition instability can reverse policies quickly when governments change. Presidential democracies with polarized politics (United States) experience policy whiplash as administrations change. Renewable-friendly administrations accelerate deployment;

hostile ones slow it. But as discussed in Chapter 6, market forces and state-level leadership increasingly maintain renewable momentum regardless of federal policy.

Petrostates with fossil fuel-dependent economies (Saudi Arabia, Russia, Venezuela) face the most difficult political challenges. Their governments derive revenue and legitimacy from fossil fuel extraction. Supporting renewable transitions requires acknowledging the end of their economic model. Yet even some petrostates are increasingly diversifying. Saudi Arabia's Vision 2030 plan seeks to provide 50% of electrical power to the kingdom from renewables by 2030.[73] Federal systems (U.S., Australia, Canada) allow regional variation where renewable-friendly states or provinces lead while others lag. This creates laboratories of democracy where successful policies spread from leaders to laggards. Federal systems are messy but resilient. National policy paralysis doesn't prevent state-level action.

Political systems absolutely affect deployment pace, but no political system can permanently stop renewable transitions driven by economics. Hostile governments can slow transitions by years, but they cannot stop them indefinitely unless willing to accept enormous economic costs that typically become politically untenable.

The Stranded Asset Trap

There is a vicious cycle facing fossil fuel interests. When they fight to slow renewable transitions, they also encourage continued fossil fuel investment to "keep up appearances". But these new investments run the risk of becoming stranded assets as renewable transitions accelerate despite opposition. Lending institutions, banks, even central governmental banks are increasingly wary of tying up their fund in fossil fuel projects. Consider a utility considering whether to build a new natural gas plant or renewable-plus-storage. If political opposition to renewables succeeds in slowing deployment, the gas plant might operate for its full 40-year expected lifetime. If renewable deployment continues accelerating, the gas plant might become economically obsolete in 15-20 years. The utility's decision depends partly on political outcomes.

This dynamic is playing out across the fossil fuel industry. Companies continued investing in oil extraction, refineries, and pipelines through the 2010s, partially because they believed political opposition would slow EV adoption and renewable deployment. Now they face situations where this infrastructure might become obsolete before recovering costs, creating enormous losses. The rational response for fossil fuel companies is to stop new investments and focus on extracting value from existing infrastructure while it is still economically viable. But this strategy requires acknowledging that the industry is in terminal decline, which is not a popular message for executives, shareholders, or employees. So, companies often continue investing based on optimistic scenarios where political opposition succeeds, creating stranded assets when it doesn't. No one wants to admit they are losing a war.

The Inevitability Argument: When Fighting Stops

Eventually, fossil fuel industries will stop fighting and start adapting. This happens when opposition becomes obviously futile and accepting reality offers better outcomes than denial. We're seeing early signs of this shift. Major oil companies are investing in renewable energy, carbon capture, and hydrogen production. Some companies are rebranding from "oil and gas companies" to "energy companies," positioning themselves for a diversified future. However, it is important to point out that many oil companies continue to act like oil companies. In 2000, British Petroleum adopted a green flower as a logo and announced their new name as "Beyond Petroleum". Despite their announcements of a pivot, they never spent more than 5% of their budget on renewables and have stepped back from most of their commitments by 2025. So, like many things surrounding the inflection point, it is frothy.

Interestingly, the timing varies by company and region. European oil majors (BP, Shell, Total) pivoted toward energy transition narratives earlier than American companies (ExxonMobil, Chevron), partly reflecting different political and social pressures. Some companies will adapt successfully; others will decline and disappear. But the industry is moving, slowly and often grudgingly, toward acknowledging renewable dominance.

The inevitability argument is pragmatic rather than moral. It says: "Renewable transitions are happening whether you fight them or not. Fighting wastes resources and makes eventual adaptation harder. Accepting reality and adapting now produces better outcomes than denying reality and fighting the inevitable." This argument becomes more persuasive as evidence of renewable inevitability accumulates.

Why Politics Ultimately Loses to Economics

The fundamental reason political opposition fails is that you cannot legislate away superior economics indefinitely in market economies. When renewables are genuinely cheaper and better than fossil fuels, markets choose renewables regardless of political preferences. Politicians can slow this through subsidies for fossil fuels, barriers to renewables, and regulatory advantages for incumbents. But these "throwing the game" actions become increasingly expensive and visible as renewable cost advantages grow. At some point, forcing utilities to choose expensive fossil fuels over cheap renewables creates political problems as voters notice high electricity rates, businesses complain about energy costs, and opposition parties campaign on eliminating inefficient subsidies.

Democratic systems have mechanisms that eventually align policy with economic reality. When policies impose obvious economic costs to benefit shrinking industries, opposition emerges from affected voters and businesses. Politicians who maintain obviously counterproductive policies get voted out. The system self-corrects, even if slowly and imperfectly. This is why even in politically hostile environments like Republican-dominated states, renewable deployment continues. When ranchers make more money leasing land for wind turbines than raising cattle, they support wind energy regardless of Republican leadership opposition. When utilities can generate electricity cheaper with solar than natural gas, they build solar regardless of political preferences. When manufacturers demand renewable electricity for supply chain requirements, politicians who block renewable deployment create problems for important constituents. Even Donald Trump's sons have invested in a data center (a bitcoin mining operation operated as American Bitcoin) that runs exclusively on wind power because it is the least expensive option. Money ultimately trumps ideology.

The bottom line is that the limiting factor on political opposition isn't morality or concern about climate change, it is purely, practical economics. You can fight economics for a while, but not forever. Eventually, the costs of swimming against economic tides become too high, and political resistance collapses or becomes irrelevant as market forces prevail.

The Fading Opposition

Fossil fuel political opposition will continue for years, possibly decades. There's too much money at stake, too many careers invested in fossil fuels, too much political infrastructure built around the industry. The fight won't end suddenly with fossil fuel interests surrendering. But opposition is weakening. The pro fossil fuel arguments are less persuasive as renewables prove themselves at scale. The financial resources are declining as the industry shrinks. The political allies are fewer as renewable constituencies grow. The tactics are recognized and less effective. Each year, fossil fuel political influence declines while renewable industries gain political strength.

By 2030, we'll likely see fossil fuel political opposition resembling the tobacco industry's current position; still fighting but marginalized, still lobbying but with diminished influence, still denying reality but persuading nobody outside their shrinking base. The fight will continue, but the outcome is no longer in doubt. The previously unstoppable oil and gas lobby will become a hit and run insurgency.

Looking Forward

The next chapter examines how the issue of grid reliability, the historical "ace in the hole" for fossil fuel interests has now become a feature of renewables.

CHAPTER 14

Grid Reliability and the Reliability Myth

O n February 15, 2021, Texas faced a catastrophic power crisis. Temperatures plummeted to record lows. Demand for heating electricity surged. And the Texas grid experienced rolling blackouts affecting millions. Over 200 people died from exposure and related causes. Political battles erupted over who was responsible. Texas Governor Greg Abbott blamed wind turbines, claiming frozen turbines caused the crisis and that this proved renewables were unreliable. He even called them "deadly". The narrative provided politically convenient headline: We told you. Renewable energy fails during extreme weather.[74]

There was only one problem: the narrative was false. Wind generation actually exceeded expectations during the crisis. The primary failures were natural gas plants and pipelines that froze because they weren't winterized. Coal plants failed. Nuclear reactors went offline. Every generation type experienced some problems, but natural gas infrastructure failures were the largest contributor to the blackout.

The Texas crisis reveals a crucial dynamic in renewable energy debates: the "reliability myth". This is the persistent claim (from fossil fuel advocates that renewable energy compromises grid reliability, despite overwhelming evidence to the contrary. This chapter examines what grid reliability means, how renewable-dominated grids achieve it, what the evidence shows about reliability with high renewable penetration, and why the reliability myth persists despite contradicting reality.

What Reliability Actually Means

Grid reliability isn't one thing; it is several distinct concepts that often get conflated:

- Available sources of generation: Having sufficient generation capacity to meet peak demand plus reserves. This is planning for worst-case scenarios, such as heat waves driving air conditioning demand, cold snaps requiring heating and simultaneous outages at multiple plants.
- Frequency stability: Maintaining constant 60 Hz (in North America) or 50 Hz (most other regions) frequency. Deviations indicate supply-demand imbalances and can damage equipment if sustained.
- Voltage stability: Maintaining appropriate voltage levels throughout the grid. Too high or too low voltage can damage equipment or cause outages.
- System resilience: Recovering quickly from disturbances such as weather events, equipment failures, unexpected demand changes. Resilient grids can withstand shocks without cascading failures.

Traditional fossil fuel grids achieve reliability through "baseload" generation (coal and nuclear plants that run constantly) plus "dispatchable" peaker gas plants that ramped up during high-demand periods. This model is familiar and well-understood, creating psychological comfort even when actual reliability was imperfect (blackouts happened regularly even in conventional grids).

Modern renewable-dominated grids achieve reliability differently. Through geographic diversity, storage, demand flexibility, sophisticated forecasting, and intelligent grid management. The methods differ, but the outcomes are comparable or better. Multiple countries now demonstrate that grids with 40% to 80% renewable electricity can match or exceed the reliability of conventional grids.[35] The proof? Countries are deploying more renewable energy than ever. This wouldn't happen if the generating resources were unreliable because the electrical consumers would not tolerate it.

The Evidence: High Renewable Grids Work Reliably

By 2025, we're not speculating about whether renewable-dominated grids can work reliably, we have extensive real-world evidence that they do:

Denmark gets 88% of its electricity from renewables (mostly wind) and maintains among the most reliable grids in Europe. Average outage duration is lower than many countries with conventional grids. The reliability metrics improved as renewable penetration increased, not worsened.[35]

Portugal reached 87% renewable electricity in 2024, combining wind, solar, and hydro. Grid reliability remains excellent. Industrial consumers in Portugal don't experience more frequent outages than those in fossil fuel-dependent countries.[35]

South Australia experienced spectacular renewable growth, going from 1% renewable electricity in 2007 to over 70% by 2024 (mostly wind and solar). Critics predicted reliability disasters. Instead, after some early challenges that were addressed through better forecasting and storage, South Australia's grid now performs as reliably as or better than when it was fossil fuel dominated.[75]

California has at times run entirely on renewable electricity during spring and early summer days (mostly solar). The grid handles these periods without reliability problems. The challenges California faces relate primarily to transmission constraints and late-afternoon ramping (when solar drops and demand rises), not fundamental unreliability of renewables.[76]

Germany integrated renewables to over 50% of electricity generation while maintaining industrial competitiveness and grid reliability. German manufacturing, among the world's most sophisticated and demanding, operates on a renewable-dominated grid without reliability compromises.[35]

There is a consistent pattern. Regions that reach high renewable penetration always experience some turbulence as they learn to manage variable generation but then achieve reliability comparable to or better than conventional grids once management systems mature. There's no evidence that high renewable penetration inherently compromises reliability when properly managed.

What Actually Failed in Texas 2021

Returning to the Texas crisis clarifies what reliability problems look like in the real world, and they are not inherent to renewables. The primary failure was inadequate winterization. Natural gas plants and pipelines lacked the optional equipment for operations in extreme cold and when temperatures dropped, equipment froze and failed. It is a fact that all types of electrical generation equipment works without issue in the frigid Northern regions of the world. The fact that the Texas grid was not winterized was a choice related to saving money. This wasn't a renewable energy problem; this was a conscious choice made by utility executives in Texas to improve their short-term profitability. Lastly, Texas's grid operates independently from other regions, preventing imports when local generation fails.[74]

Wind generation did drop below maximum levels (ice on blades reduces efficiency), but wind still generated more electricity than expected for that time of year. Wind was never intended to provide most winter capacity, in the Texas grid, that was the role of natural gas and coal. But fossil fuel generation failed catastrophically, leaving wind to carry more load than planned.

Nuclear power, a cornerstone of the supposed "reliable baseload", also failed. One of Texas's two nuclear plants went offline due to frozen equipment. So much for nuclear reliability during extreme weather.

The factual lessons from Texas 2021:

- Independence has costs: Texas's isolated grid prevented importing electricity from unaffected regions. Interconnection provides resilience.
- Infrastructure must be weatherized: Whether fossil fuel or renewable, equipment must be designed for local climate extremes, not just averages.
- Diverse generation helps: Texas over-relied on natural gas. More diversity (including more wind, solar, and storage) could have reduced the crisis severity.
- Market design matters: Texas's isolated market created perverse incentives where operators didn't invest adequately in weatherization. Profits took precedence over reliability.

None of these lessons suggest renewables compromise reliability. If anything, the crisis showed that fossil fuel infrastructure can be just as vulnerable to extreme weather as renewable infrastructure and that grid design, weatherization, and diverse generation matter more than generation technology type.

The Baseload Myth

Perhaps the most persistent reliability myth is that grids need "baseload" generation (coal, gas or nuclear plants running constantly) for reliability. This claim reflects 1930's thinking that a static network with constant capacity is the safe engineering approach. Yet, many countries have proven that with modern customers, flexibility, even to the point of putting batteries on baseload grids, is the optimum solution.

The baseload concept came from an era when demand was relatively predictable, and generation was inflexible. Coal and nuclear plants were expensive to build but cheap to operate, and they couldn't ramp up or down easily. The solution was to run them constantly at full capacity ("baseload") and use flexible natural gas plants for varying demand.

But high-renewable grids don't have predictable constant load for baseload plants. Solar floods the grid midday; wind varies with weather patterns. What modern grids need is flexible generation that can quickly ramp up when renewables are low and ramp down when they are abundant. Traditional baseload plants are actually problematic in renewable-dominated grids because they cannot adjust to renewable variations. Storage, demand response, and fast-responding natural gas plants provide flexibility better than baseload. A battery system can ramp from zero to full output in milliseconds. A combined-cycle natural gas plant can ramp significantly in minutes. A fuel celled powered microgrid switches on instantaneously. On the other hand, coal plants take hours. Nuclear plants struggle to ramp at all. For renewable integration, flexibility matters far more than baseload.

Countries that have integrated the highest renewable percentages (Denmark, Portugal, Ireland, Uruguay) have little or no baseload generation. They manage reliability through combinations of flexible resources: hydro (dispatchable), gas plants (quick ramping), batteries (instant response), interconnections (import/export flexibility), and demand response. They've proven baseload isn't necessary for reliability. The baseload myth persists because it is emotionally compelling (constant generation running 24/7 feels reliable) and it is backed by powerful interests that want to sell oil and coal. But actual grid reliability comes from matching supply and demand every second, which requires flexibility, not constancy.

The Redundancy Advantage

It's an ironic twist that renewable generation can provide more redundancy and resilience than centralized fossil fuel generation. A conventional grid depends on a relatively small number of large power plants. If one major plant fails unexpectedly (equipment breakdown, fuel supply interruption, operator error, hurricanes, etc.), the grid loses significant capacity instantly. This creates reliability risks that require extensive planning and backup capacity. The "all your eggs in a few baskets" problem.

A renewable-dominated grid with thousands of distributed generators has natural redundancy. If one solar farm goes offline, thousands of others continue generating. If some wind turbines are down for maintenance, thousands more operate normally. No single failure significantly affects total generation. What is happening in power generation today is like the move from large "mainframe" computers to distributed, desktop computing. One could say that the ultimate configuration for both data services and the electrical power grid is a topology similar to cloud computing.

The Reliability Paradox

As renewable penetration increases and reliability is maintained or improves, critics don't acknowledge they were wrong, they move the goalposts.

- When renewables were 10% of generation: "They cannot work beyond 20% without compromising reliability."
- When renewables hit 30%: "They cannot exceed 50% reliably."
- When Denmark reaches 88%: "Denmark is small and connected to neighbors. That approach would not work at larger scale."
- When South Australia is close to 100% renewables: "That's one small region. It would not work for entire countries."
- When Germany, Spain, and Portugal demonstrate high renewable penetration at national scale: "They still rely on interconnections, it wouldn't work for isolated grids."

The goalposts keep moving because the reliability critique was never about actual evidence. It was about selling more fossil fuels. No amount of real-world success will convince critics who are ideologically or financially invested in renewables failing. This matters politically. Public understanding lags reality by years. Many people still believe "renewables cannot provide reliable electricity" despite overwhelming evidence to the contrary, because the myth is repeated constantly while successes are underreported. Fighting this misinformation requires not just demonstrating reliability, but actively publicizing successes to counter persistent myths.

The Actual Threat to Reliability

If renewables don't inherently threaten reliability, what does? Several factors genuinely compromise grid reliability:

- Inadequate transmission capacity: Renewable generation concentrated in specific regions (solar in deserts, wind in plains) requires transmission to move power to demand centers. Insufficient transmission capacity creates bottlenecks that compromise reliability regardless of how the power is generated.
- Underinvestment in all grid infrastructure: Aging transformers, inadequate substations, deferred tree trimming and outdated control systems. Grids need continuous investment in infrastructure, and deferred maintenance creates reliability risks. The exploding demand across the planet is exposing underfunded grids.
- Extreme weather: Climate change is increasing frequency and severity of heat waves, cold snaps, hurricanes, and wildfires. These events stress grids regardless of generation type. Better weatherization, diverse means of generating power and resilient design are essential.
- Cyberattacks: Modern interconnected grids face cybersecurity threats. A successful attack on grid control systems could cause widespread blackouts regardless of generation technology.
- Poor market design: Markets that don't adequately value reliability or flexibility create perverse incentives. Texas's 2021 crisis partly resulted from market designs that didn't incentivize weatherization or adequate reserve capacity.
- Rapid load growth: Data centers, EV charging, electrification of heating all increases electricity demand. Grids must scale to

meet exploding electrical usage growth, and insufficient capacity creates reliability risks.

Notice what's missing from this list: renewable generation as a source of electricity. The genuine threats to reliability are infrastructure, planning, extreme weather, and market design, not whether electricity comes from solar/wind versus coal/gas.

Moving Forward on Reliability

This chapter's key message is this: The reliability myth is just a myth. Real-world evidence from multiple countries and regions demonstrates conclusively that renewable-dominated grids can operate reliably. Critics who claim otherwise are ignoring evidence, moving goalposts, or deliberately spreading misinformation to protect fossil fuel interests.

The lights stay on. The grid stays stable. Industrial processes run smoothly. All with renewable-dominated electricity. This has been proven repeatedly, at scale, across diverse regions and grid configurations. The reliability debate is over and renewables have won.

Looking Forward

The renewable revolution requires new types of materials, including metals and minerals that previous were used, but in small quantities. One of the greatest threats to the surge of renewables is securing a pipeline of the necessary raw materials or designing technology with alternate chemistries.

CHAPTER 15

Raw Materials and Supply Chains

In 2021, an anti-EV viral graphic circulated on social media showing the minerals required to build one electric vehicle battery: pounds of lithium, cobalt, nickel, manganese, copper, and graphite. The message: "This is what you have to mine for one battery. But sure, EVs are better for the environment."

The graphic wasn't wrong about the materials. But it was deeply misleading about the implications for the environment. Yes, batteries require mining. But so does every gasoline powered vehicle. Building gas and diesel vehicles requires enormous quantities of steel, aluminum, and plastics (all from mining or oil extraction), plus the oil for fuel itself requires massive extraction infrastructure and continuous fuel extraction for the vehicle's entire life (don't forget there are no oil spills or air pollution in the EV world). The relevant question isn't "do batteries require mining?" but "do they require more mining and drilling than alternatives, and are the materials available at scale?" It is also important to point out that EV batteries manufactured in 2025 are 99% recyclable, leading to a circular economy that relies less and less on mining each year.[77]

This chapter examines the raw material and supply chain challenges facing renewable energy transitions. Which materials are truly constrained, where supply chain vulnerabilities exist, how recycling and substitution address concerns, and why material constraints won't stop renewable transitions even if they create temporary bottlenecks. And more importantly, we'll compare these requirements to similar needs for similar fossil fuel uses.

The Critical Materials: What Actually Matters

Renewable energy and storage technologies require specific materials in quantities that raise questions about global supply. The key materials fall into several categories:

- Lithium (for lithium-ion batteries): A material that is used in current EV batteries and grid storage. Global lithium resources are abundant, and estimates suggest sufficient reserves for billions of EVs. The constraint isn't total availability but production capacity. Current production is roughly 500,000 tons annually; meeting projected EV demand requires increasing to 3-4 million tons by 2030. This is achievable with investment, but it requires expanding mines and refineries significantly. Keep in mind that new battery chemistries, including solid state batteries for EVs and sodium-ion batteries for the grid, may obviate the need for ramping up lithium mining.

- Cobalt (battery cathodes): More problematic than lithium. About 70% of cobalt comes from the Democratic Republic of Congo, often from mines with serious labor and environmental concerns including child labor. This concentration creates supply chain vulnerabilities and ethical concerns. However, battery technology is evolving to reduce or eliminate cobalt. Newer lithium-iron-phosphate (LFP) batteries use no cobalt, and cobalt content in traditional batteries has dropped significantly (from 20% a decade ago to often under 5% today). The search for "easier" battery chemistry is global and has significant financial backing. The cobalt problem should be temporary.

- Rare earth elements (magnets in wind turbines and EV motors): Despite the name, rare earth elements aren't actually rare. They are relatively abundant in Earth's crust, but don't exist naturally in pure forms. The challenge is that China controls roughly 60-70%

of rare earth mining and 85-90% of rare earth processing. This creates supply chain concentration and geopolitical concerns. However, rare earth deposits exist globally; China's dominance stems from being willing to tolerate environmental impacts from refining (which is a nasty business, but manageable with proper regulations).

- Copper (wiring, motors, transformers): Renewable energy systems use more copper than fossil fuel systems (solar panels, wind turbines, batteries, EV motors all require significant copper). Global copper reserves are adequate for projected renewable transitions, but expanding mining and refining capacity is necessary. Copper recycling is well-established and will provide increasing supply. Copper is not a constraint.
- Silicon (solar panels): Extremely abundant. Silicon is one of the most common elements in Earth's crust (sand is mostly silicon dioxide). The material constraint isn't an issue; manufacturing capacity is the constraint, but this easily scales with investment.
- Nickel (battery cathodes, stainless steel): Abundant globally with reserves in many countries. Nickel supply can scale to meet battery demand, though additional mining investment is needed.

Upon review, for most critical materials, the constraint is production capacity, not absolute availability of raw materials. This is a solvable problem with investment and time. Markets respond to scarcity with higher prices that encourage expanded production. Some materials (particularly cobalt from Congo) have concentration risks requiring supply chain diversification, but this is happening as industries recognize vulnerabilities, and alternate chemistries can be brought into production processes to reduce the reliance on cobalt. Once mined, almost all metals are easily recycled, leading to a circular economy in the future.

The Recycling Solution

It is important to understand a fundamental difference between fossil fuels and renewable fuels. Fossil fuels are finite, once removed and burned, they are gone forever. There is no recycling in the fossil fuel extraction industry. Renewables, on the other hand, are a manufactured technology that can be created and recycled. The "fuel" for renewables are natural forces that are endless and will never be exhausted (a huge part of the renewable cost advantage comes from the fact that renewables have no marginal, or operating, cost). Fossil fuels can serve humanity for a few centuries. Renewables can serve humanity for hundreds of millions of years.

Battery recycling is becoming a major industry. Lithium-ion batteries retain roughly 70% to 80% of their capacity after 8 to 10 years in EVs. These "end-of-life" EV batteries still have plenty of capacity for stationary storage applications and they can be repurposed for grid storage where weight doesn't matter, and slightly reduced capacity is acceptable. After another decade of stationary use, the batteries get recycled, recovering lithium, cobalt, nickel, and other materials for new batteries. An example of this is recycling start up Redwood Materials. They worked with data center constructor Crusoe to build a data center in Nevada that is completely off the grid. The operation uses a solar array to power the data center while simultaneously charging hundreds of used EV batteries. When the sun goes down, the network of EV batteries takes over and provides electricity. It is a fascinating example of the flexibility of electrical and renewable systems to adapt.[78]

Current lithium-ion recycling recovers 95%+ of cobalt and nickel, 80-90% of lithium. As recycling technology improves and battery chemistry standardizes, recovery rates will increase. By 2040, recycling will provide a significant fraction of battery materials, reducing dependence on primary mining.

Solar panel recycling is less mature but advancing rapidly. Silicon, aluminum, glass, and copper from old panels can all be recovered.

As the first wave of solar panels from the 2000s reaches end-of-life (25–30-year lifespan), recycling infrastructure is scaling up. Europe is leading with regulations requiring solar panel recycling.

Wind turbine recycling presents challenges (blades are difficult to recycle), but solutions are emerging. Newer turbines use more recyclable materials, and companies are developing processes to break down blade composites for reuse. Steel towers and copper wiring are easily recycled. The first recyclable wind turbine blades are now in use.

As renewable deployment scales, recycling will provide a meaningful portion of the raw materials, creating a circular economy that reduces primary mining requirements. This is fundamentally different from fossil fuels where you must continuously extract new material to burn. The use of materials in renewables and batteries is infinite. Fossil fuels are finite.

Material Substitution: Technology Adapts

When materials become scarce or expensive, technology adapts through substitution. This is already happening in batteries:

- Cobalt reduction: Battery chemistries have evolved to use less cobalt. Lithium-iron-phosphate (LFP) batteries (increasingly popular in EVs and grid storage) contain zero cobalt. Nickel-cobalt-manganese (NCM) batteries have reduced cobalt content from 30%+ a decade ago to under 5% in many current formulations. Future solid-state batteries may eliminate cobalt entirely.
- Sodium-ion batteries use sodium instead of lithium. Sodium is one of the most abundant elements on Earth (it is in salt water). While sodium-ion batteries have lower energy density than lithium-ion, they are suitable for stationary storage where weight doesn't matter. Chinese manufacturers are already producing sodium-ion batteries commercially at prices below lithium-ion.
- Alternative magnets: Responding to rare earth supply concerns, companies are developing motors and generators using fewer or

no rare earth elements. Some newer wind turbines use induction generators that don't require rare earth magnets. EV motor designs are evolving to reduce rare earth content.

The general principle is that when materials become expensive or scarce, markets incentivize innovation that reduces dependence on those materials. This has happened throughout industrial history (whale oil to petroleum, natural rubber to synthetic, etc.). There's no reason to expect different outcomes for renewable energy materials.

Supply Chain Diversification: Reducing Concentration Risk

The biggest supply chain vulnerability isn't material scarcity; it is geographic concentration. When one or two countries dominate production of critical materials, supply can be disrupted by geopolitical conflicts, trade disputes, or domestic instability.

China's dominance in solar panel manufacturing (80% global share), battery production (75%+ global share), rare earth processing (85% to 90%), and EV manufacturing creates concentration risks that concern Western policymakers. If geopolitical tensions escalate, supply could be disrupted or weaponized.

This concern is driving major efforts to diversify supply chains. The Inflation Reduction Act (U.S.) included substantial incentives for domestic battery manufacturing, solar production, and critical mineral processing. Dozens of factories are being built in the United States to reduce dependence on Chinese manufacturing, but again, it needs to be pointed out that the 2024 Trump Administration is working around the clock to roll back most of these programs, all of which were designed to help US manufacturing compete with China.

Europe's strategic autonomy initiatives aim to build European capacity in batteries, solar, and critical materials processing. The European Battery Alliance coordinates investments across the continent. "Friendshoring" strategies seek to build supply chains among allied countries. Australia, Canada, Chile (lithium mining),

Indonesia (nickel), and others are positioning as alternative suppliers to reduce Chinese dominance.

Technology transfer from China to other regions is happening as companies seek to serve local markets and avoid geopolitical risks. Chinese battery manufacturers are building plants in Europe and North America; solar manufacturers are expanding beyond China.

The Comparison with Fossil Fuel Supply Chains

It is worth comparing renewable material challenges with fossil fuel supply chains, which have their own vulnerabilities:

- Oil and gas supply is concentrated in politically unstable regions (Middle East, Russia, Venezuela, Nigeria). Disruptions have caused repeated energy crises, wars, and economic shocks. Oil dependence has driven foreign policy disasters, funded authoritarian regimes, and created massive geopolitical complications. There is no sun or wind equivalent to an "oil crisis".

- Coal supply chains require continuous extraction and transportation. Rail infrastructure, ports, and ships must operate constantly. Disruptions to transportation immediately affect power plants. By contrast, solar panels and wind turbines, once installed, require no fuel supply chains for decades. Coal is a dirty material. Studies have shown that up to 65,000 pounds of coal dust are put into the environment for each unit train trip from the mine to a port or a coal fired generation plant.

- Fossil fuel infrastructure (refineries, pipelines, LNG terminals) is vulnerable to sabotage, natural disasters, and aging. The February 2021 Texas crisis was partly caused by frozen natural gas pipelines, which is a supply chain failure, not generation failure.

- Dependence on hostile suppliers. European dependence on Russian natural gas funded Russia's military and created vulnerability that Russia exploited in the Ukraine invasion. This geopolitical disaster stemmed directly from fossil fuel supply chain dependence.

When considering whether renewable material supply chains are "vulnerable," the relevant comparison isn't to some idealized supply chain without risks, it is to actual fossil fuel supply chains with demonstrated vulnerabilities that have caused wars, economic crises, and geopolitical disasters. Renewable supply chains aren't risk-free, but they are arguably less risky and absolutely less environmentally destructive than fossil fuel supply chains.

The Timeline Question: Can Supply Scale Fast Enough?

There is a legitimate concern that even if materials are ultimately available and supply chains can diversify, can production scale fast enough to meet aggressive renewable deployment targets?

Current lithium production is roughly 500,000 tons annually. Meeting projected EV and storage demand requires 3 to 4 million tons by 2030. That's 6x to 8x current production in less than a decade. Can it scale that fast? Historical precedent suggests yes. Oil production increased 10-fold from 1900 to 1930. Copper production tripled from 1980-2020. Silicon production for solar panels increased 50-fold from 2000- 2020. When markets signal strong demand and high prices, mining and refining capacity scales remarkably quickly. Multiple new lithium mines are under development globally. Processing capacity is expanding. Battery manufacturers are signing long-term supply contracts, providing certainty that enables mining investment. Market signals are working as illustrated by the fact that high lithium prices (they spiked in 2021-2022) incentivized massive investment in expanded production.

The situation is similar for other critical materials. Nickel production is increasing to meet battery demand. Rare earth mining is expanding outside China and in late 2025, a significant deposit was identified in Central Utah. Copper production is growing (more slowly than desired but growing). The supply response is happening and free markets are working.

Will there be temporary shortages and price spikes? Probably. Markets don't scale in a linear fashion. They overshoot and undershoot, creating volatility. But temporary constraints don't stop long-term trends. When lithium prices spike, more mines become economically viable, production expands, and prices eventually stabilize. The market mechanism works even if it is volatile.

The consensus among materials analysts is that supply can scale adequately to meet renewable energy demand through 2030-2040, though there will be periods of tightness and high prices that create economic signals for continued supply expansion. Material constraints will slow deployment in some periods and increase costs, but they won't fundamentally prevent renewable transitions.

The Bottom Line on Materials

Material and supply chain concerns about renewable transitions are legitimate but manageable. They'll create periods of constraint, price volatility, and supply chain disruptions. They'll require continued investment in mining, processing, recycling, and diversification. They'll need strong environmental and labor standards to minimize impacts.

Looking Forward

With Part IV complete, we've addressed the major headwinds facing renewable transitions: political opposition, reliability myths, and material constraints. None of these obstacles are trivial, but none are sufficient to stop transitions driven by overwhelming economic fundamentals.

Part V (The Future is a Choice) paints a picture of a world where renewable fuels an electrified life. It also illustrates that while the eventual victory of renewables is inevitable, the timing is not. Millions of choices by governments, corporations and individuals will decide how quick the renewable transition takes place.

THE FUTURE IS
A CHOICE

Chapter 16

Vision 2050

It is 6:30 AM on a Tuesday in 2050. Your home's intelligent system has been communicating with the grid all night, charging your electric vehicle during the hours when wind power flooded the system with cheap electricity. Your house battery topped off just before dawn when solar farms across three time zones began their daily surge. The coffee maker started at exactly the moment when your personal energy cost hit its lowest point of the day. You didn't program any of this. It simply happened, invisibly orchestrated by systems that treat energy like information: abundant, distributable, and optimized in real-time.

This isn't science fiction. Every technology in that morning routine exists today. By 2050, they are simply ubiquitous, refined, and as unremarkable as running water.

The Silent City

You step outside into a city transformed by the absence of sound. The background roar of combustion engines that defined urban life for over a century has vanished. A delivery truck glides past on electric motors, its only sound the gentle hum of tires on pavement. The bus approaching the corner moves silently without the rasp of diesel engine or the cloud of black smoke. Parents push strollers down sidewalks no longer bordered by exhaust-belching traffic. Children play in parks where the air doesn't carry the acrid taste of emissions.

This silence isn't empty. It is filled with bird songs that city dwellers of 2025 forgot were possible in urban centers. The dawn chorus has returned to cities worldwide as air quality improved and noise pollution diminished. Studies from the 2040s showed that urban wildlife populations rebounded faster than biologists predicted, not because of dedicated conservation efforts, but as a side effect of eliminating combustion from transportation. The morning commute reveals the full scope of transportation's transformation. Your electric vehicle, purchased in 2047, has never seen a gas station because gas stations have largely ceased to exist in urban centers. The last ones closed in the early 2040s, victims of economics rather than legislation. Why maintain expensive underground tanks and safety systems for a fuel that fewer than five percent of vehicles use?

But your EV is more than transportation. Plugged in at your workplace, it becomes part of a virtual power plant, one of millions of mobile batteries that stabilize the grid. During today's peak demand period, your car will sell electricity back to the system, earning you enough to cover your charging costs. You'll never notice the transaction. By the time you leave work, the car will have recharged during the mid-afternoon solar surge when electricity prices briefly went negative because supply so dramatically exceeded demand.

High-speed electric rail has restructured how people think about distance. The 300-mile journey that once meant a day of driving or an expensive plane ticket now takes ninety minutes on trains powered by overhead lines connected to solar farms stretched across the route. Business travelers in 2050 marvel at old videos of airport security lines and delays, the same way people in 2025 looked back at multi-day stagecoach journeys. For distances under 500 miles, luxurious high-speed trains have recaptured the market share they lost to automobiles and aircraft over the previous century.

Aviation has split into three worlds. Silent, electric urban air taxis quietly move passengers around town, above the traffic. Short-haul flights under 800 miles increasingly use electric aircraft, their batteries charged during overnight hours when wind power dominates the grid. These planes carry fewer passengers than the jumbo jets of the past, but they fly quietly, more frequently, land at smaller airports, and cost a fraction as much to operate. For longer distances, sustainable aviation fuel derived from renewable hydrogen and captured carbon has replaced petroleum-based jet fuel.

Shipping, the last holdout of fossil fuels, has undergone its own transformation. Container ships crossing oceans now run on green hydrogen or ammonia produced by massive offshore wind installations. The largest vessels have hulls designed for efficiency rather than speed, their solar panel-covered decks generating auxiliary power for onboard systems. Port cities that once choked on diesel emissions from idling trucks and ships now breathe air that rivals rural regions.

Buildings That Think

The office tower where you work was completed in 2048, but it would have been impossible to build a decade earlier. Every south-facing surface is covered with building-integrated photovoltaics that look like ordinary glass but generate enough electricity to power the entire building on sunny days with surplus to spare. The windows themselves are smart, their tinting adjusting moment by moment to balance natural light with cooling loads. On a mild day like today, the building needs almost no external energy.

What the building doesn't generate, it draws from the neighborhood microgrid, a collection of commercial and residential properties that share energy resources and battery storage, all backed by fuel cell "peakers". When a cloud bank passes over the district, batteries instantly compensate. When the sun returns, excess generation flows to wherever demand exists in the city, guided by AI systems that predict and respond to needs milliseconds before humans would notice a problem.

Heating and cooling, which once consumed more energy than any other building function, have been reimagined entirely. Every building constructed after 2035 uses heat pumps that move thermal energy rather than generate it through combustion. Your apartment, built in 2042, maintains perfect comfort year-round while using less than a quarter of the energy that a similar space required in 2025. On the coldest winter nights, your heat pump is three times more efficient than the gas furnaces that once dominated residential heating. It runs on clean electricity, and you haven't thought about it in years because it simply works.

Older buildings have been retrofitted with the same technology. The brownstone apartments that line the historic districts of major cities now hide modern heat pumps and insulation behind their preserved facades. By 2045, building codes in most developed nations required fossil fuel heating systems to be replaced at end of life. The transition happened faster than policy makers expected because the economics made sense. Heat pumps cost less to install than replacing an aging furnace, and they cut energy bills by sixty percent or more.

Residential rooftops that aren't covered with solar panels are covered with green roofs, living systems that cool buildings naturally, manage stormwater, and provide habitat for urban wildlife. The heat island effect that once made cities five to ten degrees hotter than surrounding countryside has been cut in half. Summer nights in urban centers are tolerable again without air conditioning running at full capacity.

Industry Transformed

The steel mill on the city's outskirts would have been unrecognizable to an industrial engineer from 2025. The blast furnaces that once burned coal around the clock have been replaced with electric arc furnaces powered by renewable electricity. The hydrogen reduction system that removes oxygen from iron ore produces water vapor instead of carbon dioxide. The mill runs around the clock, but its operations flex with electricity prices. When wind power floods the grid at 2 AM, the mill ramps up to full capacity, producing steel at costs that legacy coal-fired mills could never match.

This pattern has replicated across every heavy industry. Cement production, once responsible for eight percent of global carbon emissions, now uses electric kilns and alternative chemistries developed in the 2030s. Chemical manufacturing plants coordinate their production schedules with renewable energy availability, their processes designed for flexibility rather than constant operation. Aluminum smelters, always enormous electricity consumers, have migrated to regions with abundant renewable resources, where they serve as controllable loads that help balance grids with high renewable penetration.

Industrial heat, the stubborn challenge that kept fossil fuels entrenched long after electricity generation had cleaned up, finally succumbed to innovation. High-temperature heat pumps, thermal batteries, and green hydrogen combustion now provide the temperatures needed for manufacturing processes. The transition was slower than optimists predicted in the 2020s, but the economics became overwhelming by the 2040s. Industries that clung to natural gas found themselves at a competitive disadvantage as carbon pricing and renewable costs moved in opposite directions.

Manufacturing itself has become increasingly distributed. The combination of renewable energy abundance and advances in additive manufacturing means that products can be made closer to where they are consumed. The global supply chains that once shipped components across multiple continents are shorter and more resilient. A factory in Ohio might print components that used to come from Asia, powered by solar panels on its roof and batteries charged during overnight wind surges.

Food and Agriculture Reimagined

The farm outside the city grows food using electric tractors that charge from the farm's own solar array. Precision agriculture systems, guided by AI and satellite data, apply exactly the right amount of water and nutrients to each square meter of soil. Energy use per kilogram of food has dropped by seventy percent compared to 2025 farming methods, while yields have increased. The fossil fuels that once saturated industrial agriculture, from diesel for equipment to natural gas for fertilizer production, have been replaced with renewable electricity and green hydrogen.

Vertical farms have proliferated in urban centers, their LED lights powered by rooftop solar and grid electricity. These facilities grow leafy greens and herbs year-round using a fraction of the water and land that traditional farming required. The produce travels meters instead of miles to reach consumers, eliminating the refrigerated logistics that once consumed enormous energy. Critics in the 2020s called vertical farming too energy-intensive to scale, but they were calculating based on fossil electricity prices. By 2050, with electricity from renewables approaching zero marginal cost during surplus hours, the equation flipped entirely.

Food security has improved globally, but not because of increased production alone. The reduction in climate volatility that began in the mid-2040s, as global emissions finally plummeted, meant that farmers could count on more predictable weather patterns. Extreme droughts and floods still occur, but their frequency has stopped accelerating. The window for meaningful climate stabilization that scientists identified in the 2020s was barely met, but it was met, and agriculture is reaping the benefits.

Global Equity and Energy Access

Perhaps the most profound transformation has occurred in regions that lacked reliable electricity in 2025. Sub-Saharan Africa, which had more people without electricity than any other region at the start of the century, achieved near-universal access by 2048. This didn't happen through massive, centralized power plants and transmission line buildouts. It happened through distributed solar systems, battery storage, and micro-grids that brought power to villages and towns at a fraction of what grid extension would have cost.

A farmer in rural Kenya has the same access to reliable electricity as an office worker in Copenhagen. Her solar panels and battery system power her home, charge her electric motorcycle, and run the refrigeration unit that keeps her produce fresh for market. She never knew the experience of unreliable coal power that characterized the development path of previous generations. She leapfrogged directly to clean energy, the same way her region leapfrogged from no phones to mobile phones, skipping landlines entirely.

This pattern repeated across the developing world. India, which in 2025 was still building coal plants, became the world's third-largest renewable energy producer by 2040. Its massive solar buildout not only met surging demand but eliminated the power shortages that once plagued cities and villages alike. The air in Delhi, which had been among the most polluted in the world, became breathable again as transportation and electricity generation cleaned up simultaneously.

Energy independence has reduced geopolitical tensions in ways that seemed impossible in 2025. Nations that once depended on imported fossil fuels now generate their own clean electricity from domestic sun and wind. The petrostates that wielded enormous power through oil exports have adapted or declined. Some, like Saudi Arabia, invested their oil wealth into becoming renewable energy exporters, building massive solar installations and manufacturing capacity. Others struggled to transform their economies quickly enough and faced the same fate as coal mining regions in earlier decades.

Health Transformed

The public health benefits of the energy transition have exceeded even optimistic projections. Air pollution, which killed seven million people annually in the 2020s, has dropped to a fraction of those levels. Children born in 2050 have asthma rates sixty percent lower than children born in 2025. Life expectancy has increased by two years globally, with the gains concentrated in cities and developing nations that bore the worst pollution burdens.

The elimination of combustion from transportation and electricity generation removed fine particulate matter from urban air. By 2048, even cities with populations exceeding ten million people regularly achieved air quality levels that were considered "good" by the standards of the 2020s. Beijing, Los Angeles, and Mumbai, all infamous for their smog in the early 21st century, now have air quality comparable to mountain towns. The economic value of these health improvements is staggering. Healthcare costs related to respiratory disease have plummeted. Worker productivity has increased as pollution-related sick days have declined. The savings exceed the entire cost of the renewable energy transition by some estimates, meaning the world got cleaner energy and made money doing it, even before counting the climate benefits.

Climate-related displacement, which threatened to create hundreds of millions of refugees by mid-century according to 2020s projections, has been substantially reduced. The rapid decarbonization that accelerated in the 2030s meant that the worst climate scenarios didn't materialize. Sea level rise continues, and some coastal areas have been abandoned, but the catastrophic tipping points that scientists feared were avoided. Agriculture remains viable in regions that seemed destined for desertification. The migration flows that threatened to destabilize nations have been manageable rather than overwhelming.

The Human Experience

What's most remarkable about daily life in 2050 isn't the technology. It is how unremarkable clean energy has become. People don't think about where their electricity comes from any more than they think about where their data comes from. The systems simply work, more reliably than the fossil fuel infrastructure they replaced. Power outages are rarer in 2050 than they were in 2025, despite the predictions that renewable energy would make grids less stable.

The silence and cleanliness of cities have altered the human experience in ways both profound and subtle. Urban dwellers sleep better without the constant background noise of traffic. Children play outside more because the air doesn't irritate their lungs. Streets are lined with trees that thrive because they are no longer coated with exhaust particulates. The urban experience has become more pleasant, and property values in cities have risen relative to suburbs as people rediscover the advantages of density without its previous penalties.

The anxiety that characterized the 2020s, when climate change seemed an unstoppable catastrophe, has been replaced by cautious optimism. The generation coming of age in 2050 knows climate change as history, as something their parents and grandparents struggled with and largely overcame. They've grown up with electric vehicles and solar panels as the norm. The smell of gasoline is foreign to them, something they encounter only in museums or vintage car shows.

Energy abundance has become a reality in ways that fossil fuels never achieved. The marginal cost of renewable electricity approaches zero during surplus hours, which occur more frequently as solar and wind capacity has overbuilt relative to average demand. This has enabled applications that were previously uneconomical. Desalination plants run during these hours, producing fresh water cheaply. Data centers and industrial processes time their operations to capture ultra-cheap power for use later. The scarcity mindset around energy that dominated the 20th and early 21st centuries has given way to an abundance mindset, where the question isn't whether there's enough energy but how to use it intelligently.

Community resilience has increased as energy systems have become more distributed. A neighborhood with rooftop solar, battery storage, and microgrids can maintain power during grid disruptions. Natural disasters still occur, but their impact on energy infrastructure is less severe and recovery is faster. The centralized power plants and long transmission lines that were vulnerable to single points of failure have been supplemented by distributed systems that can island and operate independently when needed.

The economic prosperity that has accompanied the energy transition has been broadly shared, though not perfectly equitably. The renewable energy industry employs more people than the fossil fuel industry ever did at its peak. These jobs are distributed across communities rather than concentrated in extraction zones. Solar panel installation, wind turbine maintenance, battery manufacturing, and grid management have created opportunities in regions that were left behind by previous industrial transitions.

Why This Vision Is Inevitable

This portrait of 2050 might seem optimistic, even utopian, to a reader today. But it is not based on wishful thinking. Every element described here emerges from trends that are already irreversible and technologies that already exist.

The economic logic that began asserting itself in the 2020s has only grown stronger. Solar and wind power became the cheapest sources of electricity, then became cheaper still. Battery costs fell by ninety-five percent from 2010 to 2030, then fell another fifty percent over the next decade. Heat pumps became more efficient and less expensive, crossing the threshold where they made economic sense even without subsidies. Electric vehicles achieved purchase price parity with gas cars in the early 2030s, then became cheaper.

Each of these technologies' benefits from positive feedback loops. More deployment drives more manufacturing scale, which drives down costs, which drives more deployment. Wright's Law, which predicted this pattern for solar panels, applied to every clean energy technology. Costs fell faster than experts predicted, and deployment accelerated faster than governments planned.

Political resistance, which seemed formidable in the 2020s, couldn't override economics. Nations tried to slow the transition through subsidies for fossil fuels and barriers to renewables, but they merely delayed their own competitiveness. Companies that bet on continued fossil fuel dominance found themselves with stranded assets and diminished market value. Investors who ignored the transition saw their portfolios underperform. The market, more powerful than any government, chose clean energy because clean energy became the profitable choice.

The fossil fuel industry fought the transition every step of the way, but their resources were finite, and their arguments grew weaker each year. They claimed renewables were too expensive, then too unreliable, then too resource intensive. Each argument was proven wrong by reality. By the late 2030s, even oil majors had transformed themselves into energy companies, divesting from fossil fuels and investing in renewables simply to survive.

The comparison to previous energy transitions is instructive. Oil didn't replace coal because governments mandated it. It replaced coal because it was better: more energy dense, more versatile, more convenient. Renewables are replacing fossil fuels for the same fundamental reason: They are better. Cheaper, cleaner, more distributed, more reliable once properly integrated. The transition was inevitable the moment renewables became economically superior, which happened in the late 2020s. Everything since has been the working out of that economic logic.

Chapter 17

The Only Question Is Timing

The world of 2050 described in the previous chapter will exist. The outcome of the energy transition is certain, but the timing is not fixed. Therefore, the only question that remained in 2025 was how quickly the transition would arrive or how long it will be delayed. Faster adoption means lower electrical bills, a more abundant life for all, a more equitable world for more people and longer, healthier lives for humans and the broader animal kingdom. Slower adoption means more suffering from pollution and extreme weather, more money being spent on energy and not health, and a greater divide between the poor and the wealthy.

If a single country works to impede the renewable transition, in favor of the resident fossil fuel interests, that country will find itself growing less and less competitive relative to nations embracing renewables. Higher costs for electricity, slower deployment of new electrical generation resources, degraded air quality will all conspire to push that country down the list of manufacturing powers and relegate them to a declining position on the global stage.

A valid analogy for our situation today is June 7, 1944: D-Day +1 in the European theater of World War II. This date marked an inflection point in the war where Germany could no longer win. Despite German military prowess, superior weapon systems and a record of stunning earlier victories, the Germans simply had no answer for the manufacturing scale of warfare equipment produced by the Americans. The Allies were now on European soil and a horde of Russians coming from the East. But the salient issue was that the Germans were about to be crushed under an avalanche of bullets, beans and bombers.

Yet even on June 7th, the outcome of the war remained unknown in human terms. Terrible battles lay ahead: the hedgerows of France, the Battle of the Bulge, the Hurtgen Forest and Operation Market Garden. The D-Day+1 parallel to today's renewable energy transition is precise. The transition's outcome is set. Fossil fuels simply cannot win in the long run because of manufacturing scale. But the path forward will be frothy, contested and involve daily "battles" where fossil fuel interests delay, obstruct and inflict damage through delayed deployment, additional costs and bureaucratic red tape.

The timing of the renewable transition remains a variable, but also a choice and the cost of delay will be measured in human suffering. Air pollution related illnesses kill approximately seven million people annually today. If the delay in renewables and electrification (including EV cars and transport) takes an extra ten years, that is seventy million lives. The total human cost of World War II. Beyond mortality, 2.3 billion people rely on dirty energy sources for home cooking, which means a polluted living environment for families and children. If electricity was available, a large percentage of these dwellings could transition to electric cooking and a cleaner home environment.

For years, fossil fuel apologists have told us that deaths from pollution related illnesses was simply the price we had to pay for modern civilization. But that price does not need to be paid any longer. We have the opportunity to save millions of lives through the quick adoption of renewable energy.

A Thousand Points of Light

What power does an individual possess against trillion-dollar fossil fuel companies and political machines built on hydrocarbon funding? We just need to look at historical precedent. The Berlin

Wall crumbled not from explosives, but from millions of individuals recognizing that its time has passed and the future would be brighter without it. Ragtag American patriots, fighting in winter without sufficient clothing or even shoes in some cases, beat the world's most powerful military force, the British, because of a shared vision of self-determination. Slavery was once deeply entrenched around the world and was responsible for much wealth, which translates to social power. Yet, the persistent determination of common people, who had had enough, eventually freed slaves across the globe.

So, what power does the individual hold? The answer is that there are a wide range of small, daily actions, that if taken by millions of common people, can make a meaningful impact on the energy transition and bring about a better world for our children. It's not the size or the strength of the challengers, but their persistence in pursuing an idea whose time has come. Fossil fuels will not be defeated; they will just become irrelevant because of a billion actions by millions of people across the globe.

George H.W. Bush, in his acceptance speech at 1988 Republican National Convention, likened America's volunteers and civic organizations as "a thousand points of light" that could "keep America moving forward, always forward – for a better America, for an endless enduring dream and a thousand points of light". It is richly ironic to quote a famous "oil man" as inspiration for the renewable energy transition, but the sentiment of his quote captures an essential truth: Ordinary people, acting with persistence and conviction on an idea whose time has come, can accomplish extraordinary change.

The Power of One

Individual choices still matter profoundly. A person buying an electric vehicle will accelerate the transition by contributing to manufacturing scale of cars and batteries. They will bring more business to charging companies, who will need to source more electricity from renewable generation companies that will employee more people and create political power. Choosing a heat pump over a natural gas furnace creates momentum for the technology, which is noticed by financial analysts and utilities, and then politicians. Installing solar panels and battery for your home gives you a forum to tell your neighbors about being independent of rising electrical bills and being indifferent to blackouts.

The author bought his first EV in 2016 and told anyone who would listen about the joys of EV ownership. Solar panels followed in 2020. The combination of solar electricity and EV batteries provided meaningful financial benefits that were easily communicated to anyone interested. He later filmed a testimonial for the company that installed his solar panels attesting to the financial savings and the delight of driving, propelled by sunshine. While the impact of this kind of advocacy is never known, his next-door neighbor, who is very politically conservative, ended up investing in a fancy solar install on a tilt mount, a battery system and a beautiful Ford Mustang Mach-E. The author's EV was the first in the small neighborhood; now there are seven.

Even as our financial world becomes top heavy with billionaires, the common consumer still wields tremendous power. In addition to making knowledgeable purchases, people can invest and support companies that are tied to renewable and green energy. You can invest in ETFs like ICLN (iShares Global Clean Energy ETF) or TAN (a solar focused ETF). You could buy shares in NextEra, the US wind, storage and solar developer. Or Enphase that makes home solar inverters and equipment. Find a bank that finances home solar. You become a vote cast daily in favor of the world you want.

Political engagement remains a real, if diminished force. Write your elected representatives. Repeatedly. Forget the abstractions of tree-hugging, saving the whales or climate change. Demand action to reduce the cost of electricity by deploying the lowest cost solutions: renewables. Talk to your neighbors, friends and relatives. Correct misconceptions with facts. Point out that the Chinese pay half what Americans pay for electricity because of going all in on renewables. Write letters to the editor at your local newspaper. Attend rallies when they are held. Learn about utility actions in front the Public Utilities

Commission and attend hearings. The author joined the Citizen's Climate Lobby (CCL) and in that capacity has attended seminars on nuclear power, sat in on lobbying calls with Congressional representatives, participated in numerous e-mail and phone-in campaigns with Congress and had Letters to the Editor published, all with the help of CCL. A list of similar advocacy groups is in the appendix of this book.

If you rent, rather than own, you can advocate for EV chargers on apartment property, for heat pumps when renovations are due or better insulation and weatherproofing. You can drive a hybrid vehicle. Champion EV chargers for your work location or try and influence your employer to use EVs for delivery or service vehicles. Corporate decisions can be influenced by employees and line managers.

The advocacy landscape is changing. Environmental organization historically focused on blocking fossil fueldevelopment (stopping drilling, blocking pipelines and protecting wild places) can now pivot towards backing popular economic solutions. When renewable energy becomes the lowest-cost electricity, the coalition expands dramatically. Business leader, farmers, manufacturers, rural landowners, and national security advocates now align with environmentalists, not through moral conversion, but through shared economic interest. This intersection creates a "big tent" political force difficult for fossil fuel interests to oppose.

The Timing is up to You

The renewable world of 2050 will exist. This is no longer a question of whether but when, a matter of years or decades, but inevitable either way. Economic logic has spoken. The technologies have been proven. Market forces have begun the transition.

But timing matters enormously. Every year of delay costs millions of lives to preventable pollution. Every year of acceleration brings the benefits of cheap electricity, healthy air, and energy independence to billions of people sooner. The difference between a ten-year transition and a twenty-year transition is measured in human suffering that could have been avoided.

You face a choice not about the outcome but about your participation. Will you lead this transition, benefit from early action, and shape the world taking form? Or will you be shaped by forces that no longer need your participation or blessing? The fossil fuel economy is ending. The renewable economy has begun. Everyone will navigate this transition. The only question is your role in it.

This book has argued that the renewable revolution is unstoppable. It is. But unstoppable does not mean automatic. It requires action. Each solar panel installed, each EV purchased, each heat pump replacing a furnace, each corporate renewable commitment, each policy barrier removed, each investment dollar redirected brings that future closer. Not metaphorically. Actually.

The renewable revolution will succeed not because humanity became wiser or more moral, but because humanity did what it always does: respond to incentives, create solutions to challenges, and relentlessly pursue what works. Clean energy works. And now it is the lowest-cost solution. Get behind it and push. Your children and grandchildren will inherit the world you help create, so make it a good one.

Appendix A

References

[1] IRENA, Record-Breaking Annual Growth in Renewable Power Capacity, 26 March 2025 Press Release.
https://www.irena.org/News/pressreleases/2025/Mar/Record-Breaking-Annual-Growth-in-Renewable-Power-Capacity

[2] Sustainability-directory.com, G20 Fossil Fuel Subsidies Dwarf Clean Energy Funding, Hindering Climate Goals, 11 May 2025
https://news.sustainability-directory.com/policy/g20-fossil-fuel-subsidies-dwarf-clean-energy-funding-hindering-climate-goals/

[3] IRENA, 91% of New Renewable Projects Now Cheaper Than Fossil Fuels Alternatives, 22 July 2025
https://www.irena.org/News/pressreleases/2025/Jul/91-Percent-of-New-Renewable-Projects-Now-Cheaper-Than-Fossil-Fuels-Alternatives

[4] Our World in Data, Solar panel prices have fallen by around 20% every time global capacity doubled, by Hannah Rice, 12 June 2024
https://ourworldindata.org/data-insights/solar-panel-prices-have-fallen-by-around-20-every-time-global-capacity-doubled

[5] PV Magazine, Lazard says fossil fuel costs double that of utility-scale solar, by Ryan Kennedy 12 June 2024
https://www.pv-magazine.com/2024/06/12/lazard-says-fossil-fuel-costs-double-that-of-utility-scale-solar/

[6] aboutAmazon.com, Amazon meets 100% renewable energy goal 7 years early, by Amazon Staff, 14 August 2025
https://www.aboutamazon.com/news/sustainability/amazon-renewable-energy-goal

[7] Apple 2025 Environmental Progress Report
https://www.apple.com/environment/pdf/Apple_Environmental_Progress_Report_2025.pdf

[8] Redway-Tech, How is Microsoft Achieving 1100% Renewable Energy for Data Centers by 2025, 5 March 2025
https://www.redway-tech.com/how-is-microsoft-achieving-100-renewable-energy-for-data-centers-by-2025/

[9] Visual Capitalist, The Exponential View of Solar Energy, by Jeff Desjardins, 25 June 2021
https://elements.visualcapitalist.com/the-exponential-view-of-solar-energy/

[10] Jasperplatz.com, Using Wright's Law to predict climate winners, by Jasper Platz, 2024
https://www.jasperplatz.com/post/using-wrights-law-to-pick-climate-winners

[11] The Texas Tribune, With Texas facing soaring electricity demand, the politics of energy quietly shift at the Capital, by Kayla Guo, 6 March 2025
https://www.texastribune.org/2025/03/06/texas-legislature-energy-renewables-power-grid/?utm_source=chatgpt.com

[12] MIT Technology Review, Inside the US power struggle over coal, by Casey Crownhart, 19 June 2025
https://www.technologyreview.com/2025/06/19/1119027/us-coal-power-struggle/

[13] IRENA, Sharp rise in energy transition investment, slide one of nine
https://www.irena.org/Digital-content/Digital-Story/2025/Nov/Progress-Shortfalls-and-Emerging-Opportunities-in-Energy-Transition-Investment/detail

[14] MIT News, Explaining the plummeting cost of solar power, by David L. Chandler, 20 November 2018
https://news.mit.edu/2018/explaining-dropping-solar-cost-1120

[15] Our World in Data, The price of batteries has declined by 97% in the last three decades, by Hannah Ritchie, 4 June 2021
https://ourworldindata.org/battery-price-decline

[16] Institute for Energy Economics and Financial Analysis, Drumbeat of coal plant closures to continue in 2025, by Seth Feaster, 15 April 2025
https://ieefa.org/resources/drumbeat-coal-plant-closures-continue-2025

[17] Energy Storage News, Behind the numbers: BNEF finds 40% year-on-year drop in BESS costs, by Andy Colthorpe, 5 February 2025
https://www.energy-storage.news/behind-the-numbers-bnef-finds-40-year-on-year-drop-in-bess-costs/

[18] Science Direct, Historical and prospective lithium-ion battery cost trajectories from a bottom-up production modeling perspective, by Sina Orangi, et al, 15 January 2024
https://www.sciencedirect.com/science/article/pii/S2352152X23031985

[19] Environment Texas Research & Policy Center, 5 surprising facts about renewable energy growth in Texas, by Ian Seamens and Evan Jones, 23 October 2024
https://environmentamerica.org/texas/center/articles/5-surprising-facts-about-renewable-energy-growth-in-texas/

[20] RMI, Wind and Solar Are Saving Texans $20 Million a Day, by Mark Dyson, 3 August 2022
https://rmi.org/wind-and-solar-are-saving-texans-20-million-a-day/

[21] Lazard, Levelized Cost of Energy slide deck, June 2024
https://www.lazard.com/media/xemfey0k/lazards-lcoeplus-june-2024-_vf.pdf

[22] Fieldvest, Oil and Gas Companies Investing in Renewable Energy: Fieldvest's Insights into the Transition, 3 March 2025
https://www.energyfieldinvest.com/post/oil-and-gas-companies-investing-in-renewable-energy

[23] General Motors, Sustainability, 2024 TCFD Report
https://www.gm.com/impact/sustainability

[24] Walmart.com, Walmart Accelerates Clean Energy Purchases and Investments With Nearly 1 GW of New Projects Across the U.S.
https://corporate.walmart.com/news/2024/03/26/walmart-accelerates-clean-energy-purchases-and-investments-with-nearly-1-gm-of-new-projects-across-the-us

[25] CEC.org, Renewable Energy as a Hedge Against Fuel Price Fluctuation, 2008
https://www.cec.org/files/documents/publications/2360-renewable-energy-hedge-against-fuel-price-fluctuation-en.pdf

[26] PV Magazine, Data centers lead global growth in corporate PPAs, by Bruno Brunetti and Caroline Zhu, 14 April 2025
https://pv-magazine-usa.com/2025/04/14/data-centers-lead-global-growth-in-corporate-ppas/

[27] Saudi Arabia Energy, Saudi Arabia's Roadmap for Renewable Energy Transition by 2023, by Marketing and Communication of Eurogroup Consultaing, 10 September 2025
https://saudienergyconsulting.com/insights/articles/saudi-arabia-roadmap-for-renewable-energy-transition-by-2030

[28] S&P Global, IPCC report outlines decarbonization pathways, wars of stranded assets, by Esther Whieldon, 27 April 2022
https://www.spglobal.com/sustainable1/en/insights/ipcc-report-outlines-decarbonization-pathways-warns-of-stranded-assets

[29] Oil Price.com, China Controls 80% of World's Solar Panel Supply Chain, by ZeroHedge, 9 May 2024
https://oilprice.com/Alternative-Energy/Solar-Energy/China-Controls-80-of-Worlds-Solar-Panel-Supply-Chain.html

[30] U.S. Energy Information Administration, China's solar capacity installations grew rapidly in 2024, 22 April 2025
https://www.eia.gov/todayinenergy/detail.php?id=65064

[31] Australian Broadcasting Company, Why China is becoming the world's first electrostate, by Jo Lauder, 12 August 2025
https://www.abc.net.au/news/2025-08-13/china-turns-into-electrostate-after-staggering-renewable-growth/105555850

[32] Microgrid Media, Solar Panel Price Difference: China vs. the US, by Jonas Muthoni, 14 February 2025
https://microgridmedia.com/solar-panel-price-differences/

[33] EnergyTrend, China Installed 45.7 GW New Solar PV Capacity in 2024Q1, 23 April 2024
https://www.energytrend.com/news/20240423-46621.html

[34] The New York Times, Quest for Clean Energy, China went from Copycat to Creator, by Max Bearak and Mira Rojanasakul, 14 August 2025
https://www.nytimes.com/interactive/2025/08/14/climate/china-clean-energy-patents.html

[35] Eurostat, Electricity from renewable sources reaches 47% in 2024, 19 March 2025
https://ec.europa.eu/eurostat/web/products-eurostat-news/w/ddn-20250319-1

[36] Earth.org, Wind and Solar Overtake Planet-Warming Fossil Fuels in EU Electricity Generation for First Time, by Martina Igini, 6 August 2024
https://earth.org/wind-and-solar-overtake-planet-warming-fossil-fuels-in-eu-electricity-generation-for-first-time/

[37] Franhofer ISE, German Net Power Generation in 2024: Electricity Mix Cleaner Than Ever, Press Release, 7 January 2025
https://www.ise.fraunhofer.de/en/press-media/press-releases/2025/public-electricity-generation-2024-renewable-energies-cover-more-than-60-percent-of-german-electricity-consumption-for-the-first-time.html

[38] Solarvision, Ember European Electricity Review 2024
https://solarvision.org/wp-content/uploads/2024-05-european-electricity-review2024.pdf

[39] Council of EU, 2040 climate target: Council agrees its position on a 90% emissions reduction, Press Release, 5 November 2025
https://www.consilium.europa.eu/en/press/press-releases/2025/11/05/2040-climate-target-council-agrees-its-position-on-a-90-emissions-reduction/

[40] gov.ca.gov, In historic first, California powered by two-thirds clean energy – becoming the largest economy in the world to achieve this milestone, press release, 14 July 2025
https://www.gov.ca.gov/2025/07/14/in-historic-first-california-powered-by-two-thirds-clean-energy-becoming-largest-economy-in-the-world-to-achieve-milestone/

[41] iaeenvironment.org, Iowa Wind Energy Fact Sheet, August 2023
https://www.iaeenvironment.org/webres/File/Wind%20Energy%20Fact%20Sheet%20-%202023.pdf

[42] CNBC, How Florida quietly surpassed California in solar growth, Lisa Setyon and Jeniece Pettitt, 2 August 2025
https://www.cnbc.com/2025/08/02/how-florida-quietly-surpassed-california-in-solar-growth.html

[43] AP News, Trump halts Revolution Wind project that's nearly complete off the Rhode Island shore, Isabella O'Malley, 23 August 2025
https://apnews.com/article/offshore-revolution-wind-project-stopped-trump-33214b9efb8f3f7a98c58299581bff9f

[44] AP News, India, a major user of coal power, is making large gains in clean energy adoption. Here is how, by Sibi Arasu, 31 May 2025
https://apnews.com/article/climate-change-india-renewable-solar-coal-wind-power-ffaaa2446482f0b96516045528ed690b

[45] PV Magazine, Kenya to combat rural energy access gap with over 130 solar mini-grids, by Cosmas Mwirigi, 14 March 2023
https://www.pv-magazine.com/2023/03/14/over-130-mini-grids-to-be-developed-in-kenya/

[46] Forbes, Uruguay's Renewable Charge: A Small Nation, A Big Lesson For The World, by Ken Silverstein, 19 October 2025
https://www.forbes.com/sites/kensilverstein/2025/10/19/uruguays-renewable-charge-a-small-nation-a-big-lesson-for-the-world/

[47] CleanTechnica, Morocco To Send Solar Power to Germany via 4800 km Undersea Cable, by Steve Hanley, September 2025
https://cleantechnica.com/2025/09/25/morocco-to-send-solar-power-to-germany-via-4800-km-undersea-cable/

[48] Mother Jones, Solar Has Been the World's Fastest Growing Power Source for Twenty Years Running, by Jillian Ambrose, 9 April 2025
https://www.motherjones.com/environment/2025/04/solar-farms-fastest-growing-electricity-power-source/

[49] Forbes, Australia to offer three free solar per day to millions, by Helen Clark, 4 November 2025
https://www.reuters.com/business/energy/australia-offer-three-hours-free-solar-per-day-millions-2025-11-04/

[50] Carbon Brief, Analysis: Solar surge will send coal power tumbling by 2039, IEA data reveals by Simon Evans, 18 October 2024
https://www.carbonbrief.org/analysis-solar-surge-will-send-coal-power-tumbling-by-2030-iea-data-reveals/

[51] LONGi.com, 34.6%!. Record-breaker LONGi Once Again Sets a New World Efficiency for Silicon-perovskite Tandem Solar Cells, Press Release, 17 June 2024
https://www.longi.com/en/news/2024-snec-silicon-perovskite-tandem-solar-cells-new-world-efficiency/

[52] MorningAgClips, The Rise of Agtivoltaics: Combining Solar Power and Crop Production, 20 November 2024
https://www.morningagclips.com/the-rise-of-agrivoltaics-combining-solar-power-and-crop-production/

[53] EnergyBot, How much electricity is lost in electricity transmission and distribution?
https://www.energybot.com/energy-faq/how-much-electricity-is-lost-in-electricity-transmission-and-distribution.html

[54] Wikipedia, Too Cheap to Meter
https://en.wikipedia.org/wiki/Too_cheap_to_meter

[55] Our World in Data, Total solar capacity, 18 July 2025
https://ourworldindata.org/grapher/installed-solar-pv-capacity

[56] PV Magazine, IEA forecasts over 4,000 GW of new solar by 2030, by Patrick Jowett, 10 October 2024
https://www.pv-magazine.com/2024/10/10/iea-forecasts-over-4000-gw-of-new-solar-by-2030/

[57] Rinnovabili, Wind Power hits Global Record in 2024 with 117GW of New Capacity, 2 May 2025
https://www.rinnovabili.net/business/energy/global-wind-power-record-117-gw-installed-in-2024/

[58] The Maritime Executive, As the Rest of the World retreats, China Doubles its Wind Power Targets, by Niu Yuhan, 26 October 2025
https://maritime-executive.com/editorials/as-the-rest-of-the-world-retreats-china-doubles-its-wind-power-targets

[59] Energy.gov, Energy Secretary Granholm Announces Ambitious New 30GW Offshore Wind Target by 2030, Press Release, 29 March 2021
https://www.energy.gov/articles/energy-secretary-granholm-announces-ambitious-new-30gw-offshore-wind-deployment-target

[60] Equinor, Hywind Scotland, undated
https://www.equinor.com/energy/hywind-scotland

[61] EnvironmentAmerica.org, Texas maintains national wind and solar energy dominance, Press Release, 7 May 2025
https://environmentamerica.org/texas/center/media-center/release-texas-maintains-national-wind-and-solar-energy-dominance/

[62] Electrek,.com ,Global wind capacity will more than double by 2030 (but, it's a shortfall), Michelle Lewis, 7 August 2024
https://electrek.co/2024/08/07/global-wind-capacity-2030/

[63] Electrek.com, Battery storage hits $675/MWh – a tipping point for solar, by Michelle Lewis, 13 December 2025
https://electrek.co/2025/12/12/battery-storage-hits-65-mwh-tipping-point-solar/

[64] PV Magazine, Cheapest source of fossil fuel generation is double the cost of utility-scale solar, by Ryan Kennedy, 11 June 2024
https://pv-magazine-usa.com/2024/06/11/cheapest-source-of-fossil-fuel-generation-is-double-the-cost-of-utility-scale-solar/

[65] Mewburn Ellis, Battery Report 2024: BESS surging in the "Decade of Energy Storage", by Rachel Pindar, 4 February 2025
https://www.mewburn.com/forward/battery-report-2024-bess-surging-in-the-decade-of-energy-storage

[66] French Development Enterprises, The Price of Power: What goes into Hydropower Project Costs?, 13 November 2025
https://fdehydro.com/hydropower-project-costs/

[67] SLB.com, Beyond levelized cost, what's the true value of geothermal energy?, by Vija Betanabhatla, 15 April 2024
https://www.slb.com/resource-library/insights-articles/beyond-lcoe-what's-the-true-value-of-geothermal-energy

[68] CarbonCredits.com, U.S. Geothermal Boom: Fervo Energy Leads with $462M Funding for Cape Station Project, by Saptakee S, 12 December 2025
https://carboncredits.com/u-s-geothermal-boom-fervo-energy-leads-with-462m-funding-for-cape-station-project/

[69] Journal of Petroleum Technology, Fervo and FORGE Report Breakthrough Test Results, Signaling More Progress for Enhanced Geothermal by Trent Jacobs, 16 September 2024
https://jpt.spe.org/fervo-and-forge-report-breakthrough-test-results-signaling-more-progress-for-enhanced-geothermal

[70] Think Geoenergy, Scientific approach to well design power Fervo Energy's Cape Station, by Carlo Cariaga, 7 March 2025
https://www.thinkgeoenergy.com/scientific-approach-to-well-design-powers-fervo-energys-cape-station/

[71] S&P Global, Hydrogen technology faces efficiency disadvantage in power storage race, by Tom DiChristopher, 24 June 2021
https://www.spglobal.com/market-intelligence/en/news-insights/articles/2021/6/hydrogen-technology-faces-efficiency-disadvantage-in-power-storage-race-65162028

[72] Reuters, Microgrids spread across US as Big Tech, utilities shore up power supplies, by Juliana Ennes, 3 November 2025
https://www.reuters.com/business/energy/microgrids-spread-across-us-big-tech-utilities-shore-up-power-supplies--reeii-2025-11-03/

[73] Saudi Arabia Energy, Saudi Arabia's Roadmap for Renewable Energy Transition by 2030, 10 September 2025
https://saudienergyconsulting.com/insights/articles/saudi-arabia-roadmap-for-renewable-energy-transition-by-2030

[74] The Nation, While Texans Freeze, Governor Greg Abbott Lies About the Green New Deal, by John Nichols, 18 February 2021
https://www.thenation.com/article/environment/greg-abbott-texas-green-new-deal/

[75] Reneweconomy.com.au, From zero to 100 pct renewables in just 20 years: South Australia's remarkable energy transition, by Giles Parkinson, 24 October 2024
https://reneweconomy.com.au/from-zero-to-100-pct-renewables-in-just-20-years-south-australias-remarkable-energy-transition/

[76] Yale Climate Connections, California just debunked a big myth about renewable energy, by Matt Simon (Grist), 6 February 2025
https://yaleclimateconnections.org/2025/02/california-just-debunked-a-big-myth-about-renewable-energy/

[77] Electrek, EV batteries are now more than 99% recyclable, by Joe Borras, 28 October 2025
https://electrek.co/2025/10/28/forget-the-myths-ev-batteries-are-now-more-than-99-recyclable/

[78] Latitude Media, Crusoe and Redwood Materials are powering a data center with old EV batteries, by Maeve Allsop, 26 June 2025
https://www.latitudemedia.com/news/crusoe-and-redwood-materials-are-powering-a-data-center-with-old-ev-batteries/

Glossary of Terms

Agrivoltaics - The practice of combining agriculture and solar energy generation on the same land, with solar panels elevated to allow farming underneath.

Baseload Power - Traditional term for continuous minimum power demand; increasingly obsolete concept as grids adapt to variable renewable generation with storage and demand flexibility.

Battery Energy Storage System (BESS) - Large-scale batteries that store electricity for later use, essential for grid stability with high renewable penetration.

Capacity Factor - The ratio of actual energy produced to potential maximum production if operating at full capacity continuously; typically, 20-35% for solar, 35-50% for wind, 90%+ for fossil fuels.

Carbon Capture and Storage (CCS) - Technology to capture CO_2 emissions from power plants or industrial facilities and store them underground; remains unproven economically and at scale.

Concentrating Solar Power (CSP) - Solar technology using mirrors to concentrate sunlight and generate heat for electricity; includes thermal storage capability.

Curtailment - Intentional reduction of renewable energy generation when supply exceeds demand and storage capacity; decreases as storage and grid flexibility increase.

Demand Response - Adjusting electricity consumption in response to supply conditions and price signals; critical for integrating variable renewables.

Energy Density - Amount of energy stored per unit of volume or mass; fossil fuels have high energy density; batteries are improving but remain lower.

Energy Transition - The global shift from fossil fuel-based energy systems to renewable and clean energy sources.

Feed-in Tariff (FiT) - Policy mechanism that guarantees renewable energy producers a fixed price for electricity fed into the grid; pioneered in Germany.

Flow Battery - Battery technology using liquid electrolytes; offers long duration storage but lower energy density than lithium-ion.

Gigawatt (GW) - Unit of power equal to one billion watts or one thousand megawatts; typical large power plant size.

Green Hydrogen - Hydrogen produced through electrolysis using renewable electricity; zero-carbon fuel for industry and transportation.

Grid Parity - Point at which renewable energy costs equal or fall below conventional fossil fuel electricity; achieved in most markets by 2020.

Heat Pump - Device that moves heat from one place to another using electricity. three to four times more efficient than resistance heating or combustion.

Intermittency - Variable nature of solar and wind power; addressable through storage, grid flexibility, and geographic diversity.

Kilowatt-hour (kWh) - Unit of energy; amount of electricity used by a 1,000-watt appliance in one hour; typical U.S. home uses 30 kWh/day.

Learning Curve/Rate - Pattern where costs decline by a consistent percentage for each doubling of cumulative production; solar has shown ~28% learning rate.

Levelized Cost of Electricity (LCOE) - Average cost per unit of electricity over a project's lifetime, including capital, operations, and maintenance; key metric for comparing technologies.

Lithium-ion Battery - Dominant battery chemistry for EVs and grid storage; costs declined 97% from 1991-2023.

Microgrid - Localized energy grid that can operate independently from the main grid; improves resilience and enables high renewable penetration.

Net Metering - Policy allowing customers with solar to sell excess electricity back to the grid at retail rates; controversial but important for distributed solar.

Offshore Wind - Wind turbines installed in ocean waters; higher and more consistent winds but higher costs than onshore; costs declining rapidly.

Peak Demand - Maximum electricity demand during a given period. Typically, hot summer afternoons in most grids; increasingly met by batteries.

Perovskite Solar Cells - Emerging solar technology with potential for higher efficiency and lower costs than silicon; approaching commercialization.

Photovoltaic (PV) - Technology that converts sunlight directly into electricity using semiconductor materials, dominant in solar technology.

Power Purchase Agreement (PPA) - Long-term contract to purchase electricity at a fixed price; common mechanism for corporate renewable procurement.

Pumped Hydro Storage - Largest form of grid-scale energy storage; uses excess electricity to pump water uphill for later hydroelectric generation.

Renewable Portfolio Standard (RPS) - Policy requiring utilities to source a specified percentage of electricity from renewables; common in U.S. states.

Smart Grid - Electricity grid using digital technology, sensors, and AI for real-time monitoring, optimization, and communication with end users.

Solar Panel/Module - Device containing multiple solar cells; typical residential panel produces 300-400 watts.

Solid-State Battery - Next-generation battery using solid electrolyte instead of liquid; promises higher energy density and safety; in development.

Stranded Assets - Fossil fuel infrastructure that loses value before end of useful life due to energy transition; includes power plants, pipelines, reserves.

Terawatt-hour (TWh) - One trillion watt-hours; unit for measuring large-scale electricity generation; global electricity use ~30,000 TWh/year.

Vehicle-to-Grid (V2G) - Technology allowing EVs to supply electricity back to the grid; turns vehicle fleet into distributed storage network.

Virtual Power Plant (VPP) - Network of distributed energy resources (solar, batteries, EVs) coordinated to function as a single power plant.

Watt (W) - Basic unit of power; rate of energy transfer or generation.

Wright's Law - Observation that costs decline by a constant percentage for every doubling of cumulative production; predicts renewable cost curves.

Organizations Driving the Transition

International Energy Organizations

International Energy Agency (IEA)
Paris-based intergovernmental organization providing data, analysis, and policy
recommendations
Website: iea.org

International Renewable Energy Agency (IRENA)
Intergovernmental organization supporting countries in renewable energy
transition
168 member countries
Website: irena.org

REN21 (Renewable Energy Policy Network for the 21st Century)
Global renewable energy community of actors from science, governments, NGOs,
and industry
Produces annual Renewables Global Status Report
Website: ren21.net

International Partnership for Hydrogen and Fuel Cells in the Economy (IPHE)
Collaboration among 23 countries and EU on hydrogen transition
Website: iphe.net

Research and Analysis Organizations

Rocky Mountain Institute (RMI)
Independent nonprofit focused on energy transformation
Influential corporate consulting and research
Website: rmi.org

World Resources Institute (WRI)
Global research organization working on energy, climate, and sustainable
   development
Website: wri.org

Lawrence Berkeley National Laboratory
U.S. Department of Energy national laboratory
Leading research on renewable energy integration and grid technology
Website: lbl.gov

National Renewable Energy Laboratory (NREL)
U.S. premier laboratory for renewable energy and energy efficiency research
Website: nrel.gov

Fraunhofer Institute for Solar Energy Systems ISE
Europe's largest solar energy research institute
Website: ise.fraunhofer.de

MIT Energy Initiative
Academic research center producing influential energy analysis
Website: energy.mit.edu

BloombergNEF (New Energy Finance)
Leading provider of research, data, and analysis on clean energy and technology
Website: about.bnef.com

Carbon Tracker Initiative
Independent financial think tank analyzing climate risk to energy markets
Coined term "stranded assets"
Website: carbontracker.org

Industry Associations

Solar Energy Industries Association (SEIA)
U.S. solar industry trade association
Market data and policy advocacy
Website: seia.org

American Clean Power Association (ACP)
U.S. trade association for wind, solar, and storage (formed from AWEA and others)
Website: cleanpower.org

Global Wind Energy Council (GWEC)
International trade association representing wind energy sector
Website: gwec.net

Global Solar Council (GSC)
International solar industry association
Website: globalsolarcouncil.org

Energy Storage Association (ESA)
Trade association for energy storage industry
Website: energystorage.org

SolarPower Europe
European solar industry association
Website: solarpowereurope.org

WindEurope
European wind industry association
Website: windeurope.org

China Photovoltaic Industry Association (CPIA)
Trade organization for Chinese solar manufacturers
Critical to understanding supply chain

Corporate Initiatives

RE100
Global initiative of companies committed to 100% renewable electricity
400+ members including Apple, Google, Microsoft, Amazon
Website: there100.org

The Climate Pledge
Commitment to net-zero carbon by 2040
Founded by Amazon and Global Optimism
Website: theclimatepledge.com

We Mean Business Coalition
Coalition of organizations working with companies on climate action
Website: wemeanbusinesscoalition.org

Science Based Targets initiative (SBTi)
Partnership defining and promoting best practice in emissions reduction
Website: sciencebasedtargets.org
Investment and Finance
Climate Bonds Initiative

International organization mobilizing bond markets for climate solutions
Website: climatebonds.net

Ceres
Sustainability nonprofit working with investors and companies
Website: ceres.org

Institutional Investors Group on Climate Change (IIGCC)
European forum for investor collaboration on climate change
Website: iigcc.org

Global Sustainable Investment Alliance (GSIA)
Collaboration of sustainable investment organizations
Website: gsi-alliance.org

Policy and Advocacy Organizations

Ember
Independent global energy think tank that uses data and policy to accelerate the
clean energy transition.
Website: Ember-energy.org

Clean Energy Ministerial (CEM)
High-level global forum of energy ministers promoting clean energy
Website: cleanenergyministerial.org

Mission Innovation
Global initiative of countries and investors working to accelerate clean energy
innovation
Website: mission-innovation.net

Energy Transitions Commission
Coalition of energy producers, industrial companies, technology providers,
finance, and environmental NGOs
Website: energy-transitions.org

Agora Energiewende
German think tank developing evidence-based and politically viable strategies for
energy transition
Website: agora-energiewende.de

Regional Development Organizations

African Development Bank - New Deal on Energy for Africa
Initiative to achieve universal energy access in Africa through renewables
Website: afdb.org

Asian Development Bank - Energy for All Initiative
Sustainable energy access programs across Asia-Pacific
Website: adb.org

Inter-American Development Bank - Energy Division
Clean energy investment and policy in Latin America and Caribbean
Website: iadb.org

Grid and System Operations

North American Electric Reliability Corporation (NERC)
Ensures reliability of North American bulk power system
Increasingly focused on renewable integration
Website: nerc.com

European Network of Transmission System Operators for Electricity (ENTSO-E)
Association of European electricity transmission system operators
Website: entsoe.eu

Smart Grid Interoperability Panel (SGIP)
Public-private partnership advancing smart grid interoperability
Website: sgip.org

Climate and Environmental Organizations

Intergovernmental Panel on Climate Change (IPCC)
UN body for assessing science related to climate change
Website: ipcc.ch

United Nations Framework Convention on Climate Change (UNFCCC)
International treaty on climate cooperation
Website: unfccc.int

Climate Action Tracker
Independent scientific analysis tracking government climate action
Website: climateactiontracker.org

World Wildlife Fund (WWF)
Conservation organization with major clean energy initiatives
Website: worldwildlife.org

Media and Communication Platforms

Electrek
Online publication covering EVs and electrification
Website: electrek.co

Canary Media
News publication covering clean energy transition
Website: canarymedia.com

Greentech Media (now part of Wood Mackenzie)
News and analysis on renewable energy markets
Website: greentechmedia.com

pv magazine
Global publication focused on solar photovoltaics
Website: pv-magazine.com

Recharge
News platform for renewable energy business
Website: rechargenews.com

Energy Monitor
Publication covering energy transition
Website: energymonitor.ai

Grassroots and Community Organizations

350.org
Global grassroots climate movement
Fossil fuel divestment campaigns
Website: 350.org

Citizens Climate Lobby
Grassroots advocacy for climate policy
Website: citizensclimatelobby.org
Local solar cooperatives and community choice aggregation programs

Author Biography

Rod Perry is a retired Silicon Valley executive who lives in Northern Utah. He and his wife, Sally, have two children and a clingy labradoodle. Rod enjoys lap swimming, international travel and sporting events at local universities and high schools.

Rod bought an electric car in 2016 and then added solar panels to his house in 2020. He and his wife both drive EVs today. Rod's experience with living the all-electric lifestyle awakened a curiosity about renewable energy, sustainable living and saving money while doing it.